Advance Praise for *North*

"Climate-driven migration is upon us. *North* is essential reading for anyone who wants to understand how and why people relocate when temperatures rise. With more people on the move, winners and losers have emerged. Importantly, Jesse Keenan shows how all of us can gain."

—**Alice C. Hill**, Council on Foreign Relations

"We need to live differently, in different places, as the physical effects of climate change accelerate, but who will make these decisions and pay for them is unclear—to everyone except Jesse Keenan, whose unrivaled expertise and wealth of insights are captured in this astonishing book. *North* spells out our shared opportunity as a society to make choices that will lead to better lives, while revealing the inside story of what has gone wrong so far on the policy front. We all need to listen to Keenan, and we are lucky he has written *North* for all of us."

—**Susan Crawford**, author of *Charleston: Race, Water, and the Coming Storm*

"The climate crisis—and the question of where Americans will migrate in response—is inextricably mingled with economic factors like insurance, mortgages, and banking. No one knows this material better than Keenan, who has woven it beautifully into an exploration of America's future."

—**Abrahm Lustgarten**, author of *On the Move*

"In this bold and visionary book, Jesse Keenan delivers a wake-up call on one of the most underrecognized consequences of climate instability: migration. Drawing on his unique insight across academia, government, and industry, Keenan charts a transdisciplinary path through the causes, consequences, and responses to climate-driven human mobility. This is essential reading for anyone seeking to understand how to preserve democracy and well-being in the face of the coming population upheaval."

—**Karen Chapple**, Director of the School of Cities and Professor of Geography and Planning, University of Toronto, and Professor Emerita, University of California Berkeley

"As the climate changes, America's plants and animals are shifting northward as their habitat changes; it makes perfect sense that people will, too. Jesse Keenan has spent years studying 'climigrants' and their effects on society, culture, housing, and business—and now it's America's turn. Lucky for us, the man knows how to write, how to advise, and how to persuade. Spoiler: He's surprisingly optimistic about the results."

—**David Pogue**, *CBS Sunday Morning* correspondent, and author of *How to Prepare for Climate Change*

North

Design Credit: Oliver Oglesby and Jesse M. Keenan.

North

The Future of Post-Climate America

JESSE M. KEENAN

OXFORD
UNIVERSITY PRESS

Oxford University Press is a department of the University of Oxford.
It furthers the University's objective of excellence in research, scholarship,
and education by publishing worldwide. Oxford is a registered trade mark of
Oxford University Press in the UK and in certain other countries.

Published in the United States of America by Oxford University Press
198 Madison Avenue, New York, NY 10016, United States of America.

CIP data is on file at the Library of Congress.

ISBN 9780197641613

ISBN 9780197641606 (hbk.)

DOI: 10.1093/9780197641644.001.0001

Paperback printed by Sheridan Books, Inc., United States of America

The manufacturer's authorized representative in the EU for product safety is
Oxford University Press España S.A. of Parque Empresarial San Fernando de Henares,
Avenida de Castilla, 2 – 28830 Madrid (www.oup.es/en or product.safety@oup.com).
OUP España S.A. also acts as importer into Spain of products made by the manufacturer.

For KLB and GWK.

Contents

List of Figures

List of Maps

List of Tables

Acknowledgments

I would like to thank Oma, Dad, Sue, Leslie, Bud, Pat, MM, Dave, and Steve, as well as my broader extended family. I am grateful for the support, guidance, and inspiration of many people, including Iñaki Alday, Jerold Kayden, Doug Parsons, Laeta Kalogridis, Michael Bell, Charles Waldheim, Alper Turan, Igor Linkov and our team at the U.S. Army Corps of Engineers, A. R. Siders, Cal Inman, Laurie Schoeman, Ann Kosmal, Ken Schwartz, Judy Kinnard, Rich Sorkin, Karen Chapple, Alex de Sherbinin, Greg Lindsay, Jennifer Marlon and her team at the Yale Climate Communications Center, Bill Solecki, Cynthia Rosenzweig, Radley Horton, Dan Schrag, Tor Törnqvist, Richard T. T. Forman, Liz Plater-Zyberk, Casius Pealer, Vivek Shandas, Juliette Kayyem, Beth Fussell, Eric Tate, Lauren Augustine, David Pogue, Abrahm Lustgarten, Tom Daamen, Tom Daniels, Eric Roston, Brianna Castro, Edgar Westerhof, Chris Calott, Rob Freudenberg, Oliver Milman, John Bibish IV, Valerie Nelson, Pete and Jessica, Jay Raskin, Lisa Salz, Brian O'Connor, Chase Cain, Clare Newman, David Waggnor, Diana Olick, Lizzy Mattiuzzi, Eric Chu, Nuin-Tara Key, Jim Venturi, John Sutter, the Gait Family, Emily Talen, Chris Flavelle, Marco Tedesco, the staff at Senator Sheldon Whitehouse's office, the librarians at Abington Free Library, Sun Dog Books in Seaside, Bunch of Grapes Bookstore in Vineyard Haven, and my climate students at Tulane University's School of Architecture and the Built Environment. I am particularly grateful for the work of our Duluth Research Team at the Harvard Graduate School of Design, including Alex diStefano, Don O'Keefe, Andreea Vasile Hoxha, Sam Adkisson, Jennfer Kaplan, Maura Barry-Garland, Sydney Pedigo, and Runjia Tian. Oliver Oglesby at the Harvard Graduate School of Design served as the designer and cartographer for this book.

List of Abbreviations and Acronyms

ABM	agent-based model
AI	artificial intelligence
APP	adaptation policy and planning
ARI	average return interval
ASCE	American Society of Civil Engineers
ASHRAE	American Society of Heating, Refrigerating and Air-Conditioning Engineers
BEES	Building for Environmental and Economic Sustainability (model)
B.P.	before the present
CAT models	catastrophe models
CCS	Climate Credit Score
CDC	U.S. Centers for Disease Control and Prevention
CDRZs	Community Disaster Resilience Zones
C&DW	construction and demolition waste
CEQ	White House Council on Environmental Quality
CFPB	U.S. Consumer Financial Protection Bureau
Citizens	Citizens Property Insurance Corporation (variously, Florida or Louisiana)
CJ	climate justice
CliFi	climate fiction
cm	centimeter
CONUS	contiguous United States
DHS	U.S. Department of Homeland Security
DOD	U.S. Department of Defense
DOE	U.S. Department of Energy
DOI	U.S. Department of Interior
DOT	U.S. Department of Transportation
EBA	ecosystem-based adaptation
ECOA	Equal Credit Opportunity Act
EE	energy efficiency
EIA	U.S. Energy Information Administration
EJ	environmental justice
EPA	U.S. Environmental Protection Agency
ESG	environmental, social, and governance
FBC	Florida Building Code
FEMA	Federal Emergency Management Agency
FFRMS	Federal Flood Risk Management Standard

FHFA	Federal Housing Finance Agency
FIO	Federal Insurance Office
FIRM	Flood Insurance Rate Map
FP&L	Florida Power & Light Company
ft	foot
ft^2	square foot
GAO	Government Accountability Office
GDP	gross domestic product
GHG	greenhouse gases
GIS	geographic information system
GJ	gigajoule
GSA	U.S. General Services Administration
GSE	Government-Sponsored Enterprise
HUD	U.S. Department of Housing and Urban Development
HVAC	heating, ventilation, and air conditioning
IAM	integrated assessment model
IAQ	indoor air quality
in	inch
IOM	International Organization for Migration (United Nations)
IPCC	Intergovernmental Panel on Climate Change
IRA	Inflation Reduction Act
IREZ	interregional renewable energy zones
kg	kilograms
kha	kilohectares
km^2	square kilometers
kWh/m^2	kilowatt hours per meter squared
lbs	pounds
LCA	lifecycle analysis
LCCA	lifecycle cost accounting
LIDAR	light detection and ranging
LTV	loan-to-value ratio
m	meter
m^2	square meter
mi	miles
mph	miles per hour
mps	meters per second
MRSLR	mean relative sea level rise
mt	metric ton
MTCO2e	metric tons of carbon dioxide equivalent
MMTCO2e	million metric tons of carbon dioxide equivalent
NAP	national adaptation plan
NASA	National Aeronautics and Space Administration
NBS	nature-based solutions
NCA	*U.S. National Climate Assessment*

NCA5	*Fifth National Climate Assessment*
NELM	New Economics of Labor Migration
NFIP	National Flood Insurance Program
NIST	National Institute of Standards and Technology
NSC	White House National Security Council
NYC	New York City
OMB	White House Office of Management and Budget
O&M	operations and maintenance
OTG	Orlando-Tampa-Gainesville mega-region
P&CI	property and casualty insurance
POPS	privately owned public spaces
PPD-21	Presidential Policy Directive 21: Critical Infrastructure Security and Resilience
RCPs	Representative Concentration Pathways
REAL	Resilience and Energy Assistance Loan
REIT	real estate investment trust
SEC	U.S. Securities and Exchange Commission
SEDAC	NASA's Socioeconomic Data and Applications Center
SEM	spatial equilibrium model
SHMP	state hazard mitigation plan
SLR	sea level rise (may be used interchangeably to denote mean relative sea level rise)
SME	subject matter expert
SoS	system of systems
SSA	U.S. Social Security Administration
SSPs	Shared Socioeconomic Pathways
sq mi	square miles
TCFD	Task Force on Climate-related Financial Disclosures
TDRs	transferable development rights
TED	tropical endemic disease zones
TIN	tax identification number
tn	tons (U.S. short)
TOD	transit-oriented development
UNFCC	United Nations Framework Convention on Climate Change
UNHCR	United Nations High Commissioner for Refugees
U.S.	United States of America
USACE	U.S. Army Corps of Engineers
U.S.C.	United States Code
USDA	U.S. Department of Agriculture
USGCRP	U.S. Global Change Research Program
WACC	weighted average cost of capital
W/m^2	watts per meter square
WTP	willingness to pay
WUI	wild-urban interface

Introduction

Driving North from New Orleans

This book explores how climate change is shaping the physical and demographic future of the United States. It is an exploration of an emerging geography of risk and opportunity that arises from the shifting landscape of people, economies, and ecologies. While some will seek to preserve the status quo, others will relocate as a territorial adaptation to climate impacts that are increasingly rendering parts of this country uninhabitable, too risky, or too expensive for permanent settlement.

My own story of moving *north* started when I decided to move south to New Orleans, Louisiana. As people have done throughout history, people move because they want a better life for themselves and for their families. They want a better education, more accessible healthcare, lower levels of pollution, cheaper housing, and greater connections with their community. Sometimes, they even want nicer weather. This is where any exploration of American mobility starts.

I had an uneasy feeling that moving to New Orleans would change my life. It was an almost sadistic sense that I would be on the frontlines of climate change. After receiving recognition for my work in climate adaptation research and public service, journalists often asked, Knowing what you know, why would you ever voluntarily move to New Orleans? As this book will explore, people are complicated.

In the summer of 2021, my wife and I found out that we were expecting our first child. Instantly, all my calculations about New Orleans and climate risk seemed irrelevant. As we processed our future, Hurricane Ida was brewing in the Atlantic Ocean. As the hurricane approached the Gulf Coast, we had little time to decide whether to evacuate. If you wait until it is too late, then you run the risk of being trapped in another Katrina-like inundation event. In the heat of a late August night, we decided that we needed to pack our most important belongings and evacuate the next day. I was concerned

North. Jesse M. Keenan, Oxford University Press. © Oxford University Press (2025).
DOI: 10.1093/9780197641644.003.0001

about my wife's health, as well as the health of our unborn child. Disasters have been well observed to have negative health outcomes for mothers and children alike.[1]

As we were leaving New Orleans driving north over the causeway across Lake Pontchartrain, the local radio station that had long served as a lifeline for the city's diaspora (WWOZ 90.7 FM) was playing Louis Armstrong's "Do You Know What It Means to Miss New Orleans." With storm clouds in my rearview mirror and sunshine ahead of me on the mainland, I felt guilty. I felt like I was leaving my neighbors with one less extra hand to clear debris and one less bag of rice to make jambalaya to feed the block. I felt that I had betrayed my implied oath of loyalty to the city that I loved.

We drove for hours through blinding bands of wind and rain. The car bore the weight of our possessions, and my body bore the weight of our uncertain future. On our first night as evacuees in Florida, I signed the contract for this book.[2] Over the coming weeks, we would migrate north from state to state avoiding COVID outbreaks in search of prenatal blood work and healthy food. We finally found refuge in Philadelphia—with nothing more than what we could fit into a mid-sized sedan. I was relieved that my son would not be born in New Orleans, a city destined to be overwhelmed by climate change. From one perspective, I figured that getting official records for everything from a birth certificate to school transcripts might be a challenge in the future. From another perspective, I wanted my son to be born in a place that he could return to at the end of his life and find solace in the continuity of the endurance of human progress.

How and Where People Will Live

The central thesis of this book is that people will shift how and where they will live in the face of climate change in the United States. Climate impacts and economic stresses are already pushing people from their homes and communities. At the same time, the prospects of lower levels of risk, more moderate weather, and a more financially sustainable way of life are pulling people to relocate. By internalizing a recognition of the risks (push) and opportunities (pull) of climate change, America is poised to enter an uncharted post-climate era.

This book provides evidence that this new era is rapidly approaching. Climate change is already altering the quality of the American way of life.

This book outlines how climate impacts shape vulnerabilities in populations and the built environment, and how the behaviors of households, markets, and governments are sending signals about the capacity of society to adapt. Across these perspectives, this book explores the dawning of a new era defined by the promise of a more sustainable way of life for those on the move and the peril of those left behind.

This book is not just a collection of scientific observations and projections about America's future. It is also a projection of optimism about America's capacity for decarbonization, environmental stewardship, and population mobility. This book is built on years of public service advising federal and state policy makers; interviews with public-, private-, and civic-sector stakeholders; and even random unsolicited calls and emails from strangers. What these people often share in common is an unrecognized optimism in an otherwise dark time.

Our popular imagination for what America's future could look like in the face of climate change is disproportionately shaped by a latent fear of the unknown. Scholars and the general public often fail to articulate how necessary adaptations will not only shape our daily lives, but how they will also reformulate a geography of risk and opportunity. We tend to focus on a constant projection of how climate change will destroy our lives without seeking to understand how we can advance a more sustainable future in the face of what we leave behind. Our default projections are constrained not only by a lack of imagination but also by our intellectual ordering of society's priorities for housing, human health, education, economic opportunity, and mobility.

This book explores how climate change is already shaping the future course of America for centuries to come. It is about the push and pull factors that are shaping our decisions. The original idea for this book came from a simple but flawed premise—flora and fauna are moving north (in the Northern Hemisphere), and so too will people and firms. Here, range shifting is not deterministically limited to mere ecological systems but rather to a full spectrum of chaotic pathways across a wide range of economic, environmental, and cultural geographies. The idea of moving "north" is both a literal reference to cardinal direction and also a metaphor for the relocations of those who will move in any direction to get out of harm's way. This book explores a series of behaviors, pathways, and scenarios that culminate in the proposition that America is on the verge of a great domestic climate migration (hereinafter, climigration) that may very well reshape everything from our physical landscape to our electoral politics.

There is not only an opportunity for climate migrants (hereinafter, climigrants) to build new communities, but there is also an opportunity to double down on our commitments to reduce our carbon footprint and to promote the accessibility, affordability, and sustainability of the built environment. We—as a society—have at least two major paths forward. In one scenario, climigrants who are economically mobile and have the means and resources can recreate their carbon-intensive settlements in the relentless exurban expansion of lower-risk places. Another scenario suggests a more orderly set of policies and behaviors that is sensitive to the environmental carrying capacity of a new sustainable frontier in "receiving zones," as well as the trash, pollution, and social inequities of what we leave behind in our "sending zones." Indeed, both scenarios may occur simultaneously depending on where you live and who you are. This book picks up on how these scenarios may play out for a wide variety of stakeholders. Along the way, there will be winners and losers. Understanding what you have to lose is the first step toward understanding what you have to gain.

Chapter 1 begins with an exploration of the mechanics of climigration in the context of international and domestic migration, including Sun Belt migration. This exploration identifies the emerging theories and models that seek to understand the relationship between migration, climate, and the environment. This chapter offers insight into the challenges of modeling both climate change impacts and the complex decision making that shapes the behavior of individuals, households, and firms. The chapter provides a roadmap for a more disciplined understanding of how academic theories and empirical evidence might seek to understand climigration. Peoples' decisions to move are based on a complex and messy set of push-and-pull factors allied to a range of known and unexplored considerations. For many, climate is already shaping their mobility decisions and options. For others, there is a great deal of uncertainty on the horizon.

Chapter 2 outlines the role that the built environment plays in driving climate change and the extent to which climate impacts define its underlying exposure and vulnerability. The chapter begins with insight into the various interacting systems that shape everything from material consumption to energy demand. This chapter seeks to understand how managing environmental impacts and greenhouse gas (GHG) emissions is a critical first step for advancing the sustainability of the receiving zones that people will settle in the future.

The main focus of chapter 2 is how climate impacts are likely to provide a powerful push factor in shaping peoples' decisions to migrate or relocate. The chapter outlines a range of impacts from extreme precipitation-driven flooding to the expanded ranges of tropical diseases that will directly and indirectly impact the occupants, structures, and systems of the built environment. In many cases, these impacts are already widely observed. The chapter draws on the recent work of the Intergovernmental Panel on Climate Change (IPCC) and the U.S. National Climate Assessment (NCA) to bridge the physical science with the applied, social, and health sciences to offer a vivid picture of the landscape of physical climate risk.

Chapter 3 provides an introductory survey of the key ideas behind climate adaptation science, policy, and planning. The chapter frames how everything from species to institutions are understood to prepare for and/or respond to climate change. Key concepts like resilience, transformation, and maladaptation are defined and explored along the way. The intent is to narrow this broad field of multidisciplinary science to a range of considerations that resonate within the arch of the built environment, public policy, and population mobility. The chapter explores how emerging climate adaptation policies and plans at the federal, state, and local level are and are not shaping the future pathways of adaptation in the United States. Understanding these institutional behaviors is key to understanding the prospects for managing the exposure and vulnerability of people and assets. Where these adaptation policies and plans fall short, people may have no other option but to relocate or migrate.

While chapter 3 highlights the adaptation of the public sector, chapter 4 outlines how the private sector is adapting to climate change, as evidenced by emerging signals in housing, real estate, mortgage, insurance, and financial services markets. From the discounting behaviors of buyers and sellers to the repricing of assets in the capital markets, market signals are telling an increasingly important story. For many, the increase in costs and the prospective flight of capital away from high-risk areas represents a fundamental challenge to remaining in place. The chapter highlights how politicians and governments sometimes want to mute these market signals for their own political ends. At the same time, the private sector is in a race to capture these signals and to monopolize climate information for their own profit. The resulting signals and noise challenge consumers' ability to make informed decisions about their own futures. As such, the chapter argues that it is the private sector and not the public sector that is increasingly driving adaptation decisions and options. While climate impacts may drive many people to relocate through displacement, others will be driven away by the hand of the market.

Chapter 5 explores the various actors and processes that define the sending zones from where people will be leaving. Sometimes these sending zones are merely local areas, while, in other cases, they represent significant centers of out-migration driving climigration. This chapter explores the various types of actors engaged in relocation, from people who are displaced by disasters to those who are preemptively looking to profit from climigration. This exploration extends to an outline of the various phases these actors engage in when making decisions about when and where to go. The chapter then shifts to the role of local governments in either unwinding and accelerating relocation or successfully managing the relocation of people, infrastructure, and a tax base. From a private-sector perspective, the chapter then explores how climate gentrification–driven shifts in demand are accelerating socioeconomic instability and displacement within prospective sending zones. The chapter concludes with a wide-ranging outline for the various factors that will need to be considered for successfully stewarding sending zones in the future.

Chapter 6 identifies the nature of the receiving zones where people will be relocating to, as well as the promises and perils of "climate havens." While the nomenclature and conceptual framing of climate havens is less than optimal, this chapter examines the critiques in public and scholarly discourse that give resolution to their rhetorical power. The chapter then shifts to a deeper study of the opportunities that receiving zones have to advance sustainability in the built environment, as new people and capital bring new ideas and energy for positive transformation. Across a range of typologies, this chapter identifies the models of settlement and development that may define this emerging geography of opportunity. With these opportunities also comes the obligation of ensuring that such advances in social and environmental welfare are captured by the diverse and inclusive range of people who will be on the move.

Finally, this book concludes with a work of fiction in chapter 7—climate fiction (CliFi). This CliFi chapter follows a family living "upstate" in the year 2079 and who are struggling to find a home in the context of massive climigration, social conflict, and environmental upheaval. The chapter chronicles the personal and psychological relationships that define their lives as climigrants, as well as the economic and physical systems of the built environment that shape their mobility. This fictional exploration offers insight into the dynamics and stresses that shape everything from migration decision making to peoples' emotional attachment to place. Set in a utopian and dystopian landscape of technologies and social organizations, the chapter speculates on the life of future climigrants and how the American way of life is adapted around them.

This book highlights that we cannot simply build resilience to an unsustainable way of life. Rather, we have to fundamentally transform our way of life from the building to the building blocks of society. This book offers insights into the various physical impacts, social behaviors, and public- and private-sector forces that are shaping our future capacity for transformative adaptation. We have the tools to achieve greater measures of human and environmental welfare. All is not lost. This book is an exploration—not a projection. It is intended to inspire all of us to see what is moving around us and to capture the moment, if only for the interest of the generations that come after us.

1
The Great Northern Climigration

Climate change is changing how and where people live. The central thesis of this book is that some people in the United States will adapt to climate change impacts by relocating and migrating. While some will relocate locally, others may very well be moving to northern latitudes for a more suitable climate with lower comparative levels of exposure to climate impacts. The title of this book is both a literal reference to cardinal direction and also a metaphor for the broader movement of people, economies, and ecologies that defy directional oversimplification. No one really knows how many people will be on the move, but it is very likely to be in the tens of millions of Americans, leading up to the end of this century.[1] This book is about preparing for this population shift and planning for what comes next in order to advance a more sustainable future.

This book is also about a changing American way of life and the countless adaptations in the built environment, the economy, and society that attempt to perpetuate a fragile standard of living. While nowhere is safe and there is no such thing as a true "climate haven," people and firms are going to settle and resettle in places from Maine to Minnesota in ways that are going to fundamentally alter the physical, cultural, and even political landscape. This book is also as much about the people who stay behind or relocate within the Sun Belt as it is about those who are undertaking a transcontinental migration. This population shift is likely to reshape the core fictions and facts that define many aspects of regional American culture. To understand the prospects of this domestic climate migration (climigration), it is necessary to first understand the fundamental relationships between population mobility and the environment.

Pushing the Limits of Migration and Immigration

Every year people around the world pack up their homes (or what is left of their homes) and leave for a variety of reasons, including reasons shaped

North. Jesse M. Keenan, Oxford University Press. © Oxford University Press (2025).
DOI: 10.1093/9780197641644.003.0002

by climate impacts and environmental exposure.[2] Individual, household, and community migration decisions and the study of people's mobility are incredibly complex. People's decisions to move are shaped by a variety factors.[3] Some of this reasoning is not always centrally based on a rational self-interest to advance one's wealth and well-being. People move in dialogue with and concern for their extended family and for others in their community.[4] Climate migrants (climigrants) have also been observed to employ a wide range of affective reasoning in their decisions to move. Amitav Ghosh has argued that digital technology has allowed for a proliferation of narrative fictions that cloud the complexity and challenges of transnational climigration. He noted that "Bangladeshis dream of escaping from crowded Dhaka and making a life in Finland, 'a quiet and empty country' with 'large fields and empty spaces.'"[5] In this sense, climigration may be the fulfillment of a dream. Unfortunately, life is rarely so simple or easy.

Pinpointing the role of climate change is not so easy either. People often move for a whole host of reasons—rational and irrational. Sometimes people are chasing a dream with a renewed sense of hope. In other cases, people begin to believe that investing their time, energy, and wealth in a high-risk area is simply too risky over the long term. There are always push-and-pull factors at work. What observers do know is that climate change is often directly or indirectly part of the calculus for an awful lot of people.

In an international context, environmental impacts, such as droughts impacting agricultural societies, have been observed to become points of socioeconomic instability that are exploited by a wide variety of destabilizing actors, from politicians to terrorists.[6] According to the United Nations High Commissioner for Refugees (UNHCR), "70% of refugees and asylum seekers [in 2022] fled from highly climate-vulnerable countries."[7] In 2023, just over 21 million refugees originated in places with significant climate exposure.[8] In some of the most vulnerable geographies, one early model forecasted that upward of 143 million people from sub-Saharan Africa, South Asia, and Latin America could be internally displaced because of climate change by 2050.[9] All these forecasts and projections are dependent on the rate and depth of climate impacts and adaptations.

People are already moving north in large numbers to the United States. The movement of people north to the southern border with Mexico was just over 2.1 million in 2024.[10] Only about 31% of those migrants encountered in 2024 were Mexican citizens.[11] Another 18% were from El Salvador, Guatemala, and Honduras (the Northern Triangle). A slim majority of migrants are coming from around the world, including Venezuela, Peru, Turkey, Romania, and China.[12] With or without climate-driven displacement, very few of the millions of migrants traveling north to the United States every year will ever be granted refugee status under current immigration laws and policies. In 2024, only 25,358 migrants were granted refugee status from Latin America and the Caribbean, with most of the roughly 100,000 admitted refugees originating from the Democratic Republic of Congo, Syria, and Afghanistan.[13]

Migrations from Mexico and the Northern Triangle countries are increasingly understood to have a climatic and environmental connection. Research has observed that droughts in the Northern Triangle[14] and rainfall deficits in Mexico were a strong predictor of migration.[15] Extreme heat waves in Mexico have been observed to reduce the demand for rural labor and drive out-migration to the United States.[16] Across this region, internal out-migration from rural areas where agricultural economies are hit hard by changing environmental conditions is anticipated to significantly increase urbanization, which, in turn, is anticipated to increase the risks of violence and crime, among other implications.[17] This urban stress may have an amplified effect for increasing migration to the United States.

A 2021 White House report on international climigration noted that "although most people displaced or migrating as a result of climate impacts are staying within their countries of origin, the accelerating trend of global displacement related to climate impacts is increasing cross-border movements, too, particularly where climate change interacts with conflict and violence."[18] The threat of violence is a major driver of southern border migration, and climate change is projected to only intensify these push factors. The United States was sufficiently concerned about climate-driven migration during the Biden administration that it actively integrated climate impacts within migration early-warning systems, such as the U.S. Immigration and Customs Enforcement's Biometric Identification Transnational Migration Alert Program, to support everything from targeted foreign assistance to refugee resettlement.[19]

Intelligence systems and foreign aid offer little recourse to persons seeking legal status as a prospective refugee in the United States because neither climate impacts nor natural disasters represent a qualifying status in a legal code that is primarily centered on protecting people facing unique risks from targeted persecution and violence.[20] This legal limitation highlights a big difference between a refugee and a migrant. A refugee has legal status as a protected person under domestic and international law, whereas a migrant may or may not classify as a refugee.[21] It all depends on the context. To make matters more challenging, there are no provisions in international law that recognize climate or environmental migrants as a category of lawfully protected refugees.[22] There is technically no such thing as a "climate refugee."

While there is a scenario where climate impacts drive socioeconomic instability that drives a qualifiable form of violence, persons displaced exclusively because of localized climate impacts lack legal standing. From an evidentiary point of view, the causal relationship between climate-driven resource scarcity and violence is conceptually strong but empirically limited, as such causality often "def[ies] simple and sensationalist conclusions."[23] As the lawyer Yumna Kamel from Earth Refuge noted, "It all depends on how you plead the case."[24] In this sense, some lawyers have successfully represented persons displaced by climate-attributed shocks or stresses by specifically documenting the qualifying forms of violence that often follow such events.

Defining and Modeling Climigration

Very few international climigrants are likely to ever qualify as a refugee. By the same token, one should recognize that not all climigrants are displaced persons. Although the term "climigrant" was originally coined by the Alaskan lawyer and scholar Robin Bronen to describe "forced permanent migration" from climate impacts, the terminology can be applied to a wide variety of mobility contexts, including forced and elective migrations.[25] As will be explored in chapter 5, some classes of climigrants may be preemptively mobile, wealthy, and even opportunistic. The term climigrants captures a broad class of individuals who may very well make long-distance transcontinental moves, but it also includes those who are simply relocating locally or regionally. In addition, for every climigrant there may also be someone that is a voluntary non-migrant who either chooses or has no other option but to remain in place despite the risk.[26]

The boundaries between migration and non-migration can be fuzzy, with temporary or seasonal movements representing a potentially valuable form of adaptation.[27] The impacts on demography are also not as simple as a net calculation of people moving in and out. For instance, with "demographic amplification" the composition of people coming and going matters.[28] As preeminent climate demographer Mathew Hauer and colleagues noted in the context of sea level rise (SLR), "Migration is most likely to occur in more youthful populations, [and] areas experiencing accelerated climate out-migration could face accelerated population aging."[29] In this sense, climigration should be conceptualized not just as a situation where someone moves from place A to place B. Rather, the very social fabric of places A and B will change over time, and that will, in turn, shape the lives of the residents who might be on the move by and between these places. In the context of Hauer's observations, some young people just might contemplate migration because their dating life in a town full of old people is dim.

Modeling migration is complex. Modeling climate-driven migration is very complex. The integrated modeling of climate change and climate variability is itself extremely challenging across scales of time and space. Climate impacts themselves may manifest as rapid onset events or slow chronic stresses. There are lots of emerging analytical models to estimate, project, and forecast climigration. Yet these "different models have produced results that are not always consistent with one another or robust enough to provide actionable insights into future dynamics."[30] There is a very active debate about methodologies, thresholds, motivators, inhibitors, and other factors that shape individual and household climigration behavior.

Part of the challenge of forecasting climigration, or any migration for that matter, is that model uncertainty is compounded by the fact that data on historic migration phenomena (e.g., who, where, when and why) is not great—even in the United States. As the sociologist Kerilyn Schewel noted:

> Migration models tend to focus on economic or demographic variables, but social, political, and cultural factors also play a role in determining who migrates, where they go, and the degree of choice in the migration process. Migration modelers still struggle to capture the nuanced relationship between these interacting drivers of migration in different socioeconomic contexts due to difficulties with gathering or accessing the relevant data as well as challenges isolating the relevant variables from one another theoretically and empirically.[31]

By adding the uncertainty of climate sensitivity and impacts—in terms of attribution, timing, depth, and location—to the mix, climate-sensitive migration models represent a challenging area of research. The first basic modeling approach is called an "exposure model." In this case, models overlay hazards and climate impacts onto a population map. This is useful for understanding how many people might live in an SLR zone, for instance, but it does not necessarily help one understand how many people may remain in place or relocate over time.[32] The advantage of these models for policy makers is that they are simple to interpret and the results are easy to communicate.

However, users of these models run the risk that any long-term planning based on these static snapshots of exposure might not foresee intervening behaviors or interacting and compounding risks from the hazards and impacts themselves. For example, building a large seawall to protect a population from SLR might be a waste of money if people move out of the SLR zone prior to the seawall reaching the end of its useful life. Indeed, they could have moved away owing to riverine flooding and not SLR. If this precarious population's property taxes pay for the wall, then this overly specialized infrastructure investment might not make sense.

For many decades, demographic modeling techniques were centered on gravity models, "which [posit] that the volume of migration between two locations increases with population sizes in each location and decreases with geographical distance between the locations."[33] These models try to understand the attractiveness between two places with the assumption that the farther apart that they are, the weaker the connection becomes. Gravity models are great for reproducing past shifts in population by and between countries over long periods of time, but their inherent determinism has not held up well for understanding dynamic processes and behaviors.[34] Two places close in proximity to each other may also share a similar climate exposure profile, which may cloud any attribution of migration to climate impacts.

By contrast, agent-based models (ABMs) are data-intensive models that attempt to understand behavior at a sub-national, urban, or local scale. ABMs seek to "identify or hypothesize the rules of behavior that lead to migration decisions in a context of multiple stimuli. A computer simulation then allows researchers to observe the outcome on a population of agents over time and to modify the contextual parameters."[35] Based largely on longitudinal household survey data, ABMs seek to draw inferences about the household characteristics and composition of who stays and who goes. The core debate within ABM development centers on the theory of behavior utilized. These theories range from agents being utility maximizers (e.g., unemployed move toward places with attractive wages) to psychosocial and cognitive theories based on individual attitudes, perceptions, and subjective norms and experiences often captured in surveys.[36]

Even seemingly little things like one's friends' or family's shared experience with migration can shape expected values within decision making, and this kind of information is not always well captured in surveys.[37] Some potential migrants have a subjectively strong attachment to the place that they call home.[38] This kind of data is only often captured in direct interviews and small-batch questionaries that appreciate the power of place in their design. At the end of the day, ABMs require lots of data that may not always be available, and they are largely limited to understanding short-term movements of people on a finite geographic scale.

Economists are also active in trying to model climigration. They often start with a spatial equilibrium model (SEM), wherein an "equilibrium occurs when neither households nor firms have incentive to relocate to another location, in which household utility and firm profits are equalized across space."[39] This perspective is helpful for understanding incentives and disincentives for adaptation to climate impacts, even though such models assume on some level that people and firms are freely able, willing, or capable of making trade-offs between climate risks and their own welfare and profit.[40] An increasingly active area of scholarship revolves around the framework associated with the New Economics of Labor Migration (NELM). NELM situates migration decision making in the context of a potential migrants' wider household and social relationships. By looking at wage differentials and "failures in insurance, credit and savings markets," NELM argues that household decision making is driven "to maximize expected income, but also maximize status within an embedded hierarchy, to overcome barriers to capital and credit, and to minimize risk and diversify the incomes."[41]

While this broader body of work in econometric and statistical model development has proceed without a common methodological or theoretical approach, these statistical models have contributed important empirical and theoretical insights based on real-world experiences.[42] For instance, migration research in 115 countries from 1960 to 2000 found that warming trends in middle-income countries increased the probability of out-migration to urban areas, whereas warming in poor countries resulted in a lower probability of migration likely due to "severe liquidity constraints."[43] This research did not identify the drivers or mechanisms per se, but this kind of multidisciplinary research nonetheless offered valuable insight for policy makers and urban planners.

Sociologists, anthropologists, geographers, demographers, and economists have been active in defining multidisciplinary climigration scholarship. This body of research has "investigated the entire cycle of migration, from intentions, the decision to migrate, the journey itself, [and] the consequences of climate migration."[44] Emerging climigration frameworks supporting a wide range of approaches from "aspiration capability models" to "habitability assessments" are inspiring a new generation of researchers who are seeking to bridge the science and social science.[45] This research has looked at decision making at the scale of the individual, the household, and extended social networks. The research has examined climate's direct role in pushing people out, but it has also looked at climate's indirect role in affecting conventional migration drivers, such as economic opportunity[46] and conflict and violence.[47]

This body of research has set up competing analytical perspectives in the models between environmental drivers and socially causal mechanisms. Here, "[e]nvironmental-[d]rivers model[s] will hold the social context as fixed and quantify the incremental damages of a measure of climate change, while a [s]ocial-[c]ausal model will show how damages are generated by social vulnerability and its antecedents."[48] This social-causal perspective is particularly useful for explaining how climate impacts may yield very different results by and between well-off and not so well-off exposed populations. These models have sometimes produced competing ideas and evidence for understanding the mechanisms of push and pull. As is always the case, it depends on the context.

For instance, some research has suggested that rapid onset shocks that come from extreme events lead to more out-migration, while other research has shown that such extreme events actually limit resources and increase immobility.[49] Both things can be true at the same time. Some segment of the population that has the wealth and resources might very well adapt by moving out, while other people are functionally trapped in place despite the desire to get out.[50] As will be explored in this book, this potential division between the haves and the have-nots risks exacerbating existing spatial patterns of concentrated poverty and inequality. For communities in high-risk areas facing an exodus of people and firms (sending zones), the economic implications for everything from increased utility and tax bills to decreased asset values are potentially severe. The social implications are also potentially troubling, as there may be fewer people able to provide the support in sending zones that is critical for societal functioning and civic engagement.

While there is little consensus around how to model and forecast climigration, it is increasingly recognized that climigration is nonlinear and that conventional metrics of temperature and precipitation are ill suited to capture the full range of environmental and climatic stressors. Some research has found that climate impacts, such as drought and temperature anomalies, do not hold the same power over migration as do "political violence and repression."[51] After all, climigration—at least in some form—is considered a way for people to mitigate risk and adapt to a wide range of climate impacts.[52]

To move beyond temperature and precipitation, scientists have been actively incorporating ensemble models widely used in climate science, such as Shared Socioeconomic Pathways (SSPs) and Representative Concentration Pathways (RCPs), in order to expand the horizons of what climate-driven behaviors and impacts might look like the in future.[53] RCPs are time-series-driven scenarios developed from integrated assessment models that account for greenhouse gas (GHG) emissions; changes in aerosols, active gases, and land use; the public policies that shape these concentrations; and the underlying climatic sensitivities that translate to radiant forcing and climate impacts.[54] The four RCP scenarios account for a range of outcomes, where RCP 2.6 is less than great news and RCP 8.5 is absolutely catastrophic, with RCP 4.5 and 6.0 somewhere in between. The numbers behind the acronyms denote the total amount of stabilized radiant forcing (i.e., how much heat is trapped) attributable to GHGs, and this is measured in watts per square meter (W/m^2). Beyond migration modeling, these RCP pathways have become shorthand for everything from public policy to engineering practice.

SSPs were developed as a complement to RCPs to account for socioeconomic behaviors and path dependencies under five different emissions scenarios from SSP1 (green) to SSP5 (fossil-fuel intensive).[55] Together, these pathways standardize assumptions and language that are widely utilized in policy and practice. Although their integration offers the promise of advancing climigration scholarship, it also presents a new range of analytical and computational complexities. In response, new methods and techniques grounded in machine learning have arisen. These techniques offer a computationally efficient means of handling the arbitrariness that often defines a complex array of interacting social, economic, and environmental drivers and predictors.[56]

Unfortunately, more is not always better. Policymakers and planners often have to rely on a range of working heuristics in the face of complex machine learning models they simply do not have the capacity to interrogate or fully understand.[57] As the United Nation's International Organization for Migration (IOM) noted, "Quantitative results from climate mobility models are likely not yet at a point where they can reliably inform policy decisions related to future in- or out-migration that would be relevant for matters ranging from labour availability to regional development, from border policy to migrant protection and assistance."[58]

At the end of the day, observing, estimating, and forecasting climigration is a messy business. What matters most is that someone aspires to and has the capability to migrate.[59] Policymakers might never be able to reliably forecast climigration, but they will need to prepare for the possibility that some places are going to hollow out and other places will explode with the weight of climate-driven population growth.

The End of the Era of Sun Belt Migration

The relationship between the environment, climate (not climate change), and domestic migration has been studied for generations. The geographer "[Ernst Georg] Ravenstein (1889) may have been the first to suggest that various factors, including 'an unattractive climate,' tend to push persons from one area to another area."[60] As the sociologists Elizabeth Fussell and Brianna Castro argued, "Migration theories mainly conceptualize the environment as an indirect influence on migration, operating through unemployment or wage differentials that exist between origin and destination communities or political, social, and demographic drivers. The environment is largely invisible in much of the contemporary migration literature, not because it is irrelevant, but because it is a relatively stable feature."[61] With a historically stable climate, the popular consensus has been that people in the United States do not mind the heat but they hate the cold.

Since the 1960s, people in America have been moving to the Sun Belt—the Southeast to the Southwest—in large numbers. By the 1990s, the Sun Belt population shift had not only accelerated, but it was also deeply ingrained in the popular culture. Nearly everyone from Boston to Chicago has a friend or relative living in Tampa or Atlanta. Evaluating the effects of 11 different climatic variables during this period, research found evidence that temperature, humidity, and wind velocity had robust effects on net migration outcomes under the theory that "persons tend to avoid exposure to bitter and cold winters, and excessively hot and humid summers, preferring climates between these extremes."[62]

While many scholars accepted this general shift in preferences (i.e., increase in demand) associated with leaving the cold climates of the Midwest and Northeast, others were less convinced that this could explain the totality of motivations. Harvard economist Edward Glaser and colleagues argued that early periods of Sun Belt growth could be explained by an increase in demand, as evidenced by significant increases in productivity.[63] However, later stages of growth in the Sun Belt were likely driven by significant increases in the supply of comparatively more affordable housing. As such, it was supply and not demand that fueled the accelerated growth, which "implies that there has been no increase in the willingness to pay for sun-related amenities. As such, it seems that the growth of the Sun Belt has little to do with the sun."[64] Warmer weather might have fueled an initial demand, but the cheap supply of suburban housing is what accelerated in-migration. The jobs merely followed a lower cost of living and lower predominately nonunion wages.[65]

The popular aversion to cold weather might not be as powerful as people once thought. As it turns out, research highlights that what people really prefer are places with "nice weather," and they are more than willing to move to cooler and less humid places.[66] The initial waves of migration to the Sun Belt were likely facilitated by the widespread diffusion of air conditioning technologies.[67] However, in contemporary terms, the effects of air conditioning are largely considered minimal. The exception is for households who are not wealthy enough to afford air conditioning, for whom this lack of habitability is estimated to be a potential driver of climigration in the future.[68]

Since the late 1990s, various research has highlighted that "the faster relative growth of places with cooler, less humid summers establishes that air conditioning cannot alone account for the movement toward nice weather. And the approximately equal draw of nice weather for both working-age and elderly individuals suggests that increased retirement has been a relatively unimportant source of the population shift."[69] While the Sun Belt had an initial draw for nice weather among retirees, the Midwest and the Northeast might very well be catching up for everyone else, as the winters moderate.

Separate research that sought to estimate behaviors under future (2020–2050) climate scenarios found that "households' [willingness to pay (WTP)] to avoid hotter summers is greatest in the areas that are expected to experience . . . increases in summer temperature—the South and parts of Southern California."[70] For those living in the South, the value of their present choice of location may be decreasing as temperatures, particularly summer temperatures, become more extreme. By the same token, the research found that "households in the Midwest and Northeast have lower WTPs to increase winter and reduce summer temperatures than households in the South and West."[71] For many people living in the North, a comparatively milder future climate will be fine with them. For recent transplants to the South, the extreme heat may be more than they economically bargained for.

The most robust evidence that the end of the era of Sun Belt migration might be nearing has come from research published by economists at the Federal Reserve Bank of San Francisco. In their 2024 paper, "Snow Belt to Sun Belt Migration: End of an Era?" the researchers found that, from 2010 to 2020, net migration rates were uncorrelated with mean counts for extreme heat and extreme cold days.[72] Although cold temperatures were observed to initially correlate with net out-migration and warm temperatures correlate with net in-migration going back to the 1970s, the researchers found a gradual decline in these relationship over the course of 50 years. Their findings held for both urban and suburban counties. They even found evidence that the correlations had reversed in rural counties: "Among rural counties, hotter places actually lost population over the latest decade of data relative to colder places."[73] The effects were strongest for people in their 20s and 60s, which suggests that climate is an "especially important factor for those in life stages involving long-term location choices."[74] With significant projected changes in the climate of the Sun Belt that include a greater likelihood for extreme heat, the researchers concluded that their "findings suggest [that] the 'pivoting' in the U.S. climate-migration correlation over the past 50 years is likely to continue, leading to a reversal of the 20th century Snow Belt to Sun Belt migration pattern."

These findings should be contextualized to the findings of Edward Glaeser and Kristina Tobio, who noted the significant draw of cheap housing. As highlighted in chapter 4, the costs of housing, insurance, financial services, and energy are going up in high-risk geographies, many of which are located in the Sun Belt. These increases in costs are on top of existing decreases in the supply of affordable housing and increases in housing[75] and energy cost-burdened[76] households in the Sun Belt.[77] By contrast, even among wealthier, high-skilled, and better-educated households, the initial advantage of the Sun Belt's relative wages (e.g., lower wages but cheaper cost of living) have been found to erode as people demand "higher incomes [to compensate] for the increased risk from natural disasters" and extreme heat.[78] If peoples' preferences for sunshine are waning—and even reversing in favor of cooler weather—and the initial draw of affordability is also waning, then the end of the era of Sun Belt migration may very well be on the horizon.

A Changing Environment in Historical Context

People are not the only ones (or things) moving north. One meta-analysis of more than 1,000 terrestrial species found that their ranges were moving north, in the Northern Hemisphere, by a median rate of 16.9 km (10.5 mi) per decade.[79] These species bring with them seeds, diseases, other organisms, and a range of novel and asynchronistic trophic interactions. For environmental managers, the distinction between an alien invasive species and range-shifting migrant species begins to break down in the context of climate change.[80]

Understanding how species and ecologies are adapting through phenotypic adaptation and range shifting is important for understanding not only climigration and human settlement but also a range of indirect impacts from climate change. People intuitively understand that the world will be hotter and wetter—it is already happening. What they might not understand is that pests expanding into warmer areas kill trees that fuel more intense forest fires[81] or that extended droughts kill plants and change soil composition in a manner that makes flooding more severe.[82] Given these wide-ranging indirect ecological and biophysical changes associated with climate change, one can imagine how difficult it will be to precisely forecast the climate-driven movement of people.

This is not even accounting for the near impossibility of anticipating sudden extreme events that climate change might drive or support, such as a pandemic from cross-species viral transmission.[83] For instance, the Zika virus outbreak in 2015–2016 not only altered tourism in the Caribbean,[84] but it would go on to interact with long-term drought, ongoing instability in agricultural occupations, and other socioeconomic factors to destabilize populations and drive out-migration to the United States.[85] Even in the event that climigration models could account for all these types of destabilizing events and risks, modeling the pull factors for places with ostensibly lower risks is questionable given the widespread impacts possible in the places where people end up (i.e., receiving zones). Modeling goes only so far toward understanding the dynamic push-and-pull factors associated with climate risk and migration.

Archaeology has provided robust evidence that human settlements—as part of complex social-ecological systems—are dynamically responsive to changes in the environment.[86] At the same time, history has shown that there are also limits to any civilization's adaptation to changes in the environment. Recent scholarship in archeology, history, and prehistory has argued that prior changes in climate have been a critical factor in driving the collapse of civilizations, cultures, and states. This includes everything from the fall of Egyptian Old Kingdom (c. 2700 to 2200 B.C.) during the 4.2 ka B.P. three-century long megadrought (~2200 to 1900 B.C.)[87] to the fall of Ming dynasty (1368 to 1644 A.D.) during the Little Ice Age (~1250 to 1850 A.D.).[88] Although some scholars have viewed climate change as an overly simplistic and deterministic explanation for collapse,[89] other scholars have carried forward examples of how people have endured with measures of resilience and adaptation.[90]

Societies and civilizations do not collapse overnight because of a changing environment. Planes rarely ever crash for just one reason. The archaeologist Guy Middleton studied the cases of civilizations and states from Angkor to the Western Roman Empire in the context of contemporary "collapsology" studies, and he concluded that a "collapse is primarily a social process driven by people rather than an apocalyptic disaster caused by an external shock."[91] The archaeologist Scott A. J. Johnson carried forward this social thesis by noting:

> The hubris of any society will prove to be its downfall. Today, many believe that life will go on much as it has for the past fifty years: prosperity, rapidly developing technology, and improved quality of life. Politicians are proud to say that America, for example, will endure forever. Egyptian pharaohs said the same thing about their valley kingdom, as did the ancient Maya about their rainforest cities. The Romans felt invincible within their empire. Simply saying that things will remain the same or improve, just because they have for the last few generations, is hubris.[92]

This book does not argue that the United States will collapse in the coming generations from climate change. Rather this book builds on the proposition that technological, political, and cultural hubris may serve as a barrier to supporting the kinds of adaptation necessary for managing the relocation of climigrants out of sending zones and for taking advantage of the opportunities for sustainable development in receiving zones. While the behavior of governing institutions may be uninformed by the lessons of history, individuals and households are wide awake to the clear and present dangers of climate change.

From one perspective, the relocation and climigration of people, firms, species, and ecologies will no doubt accelerate the micro-collapse of local cultures and communities. Cultural heritage and practices,[93] foods and recipes,[94] and places of spiritual resonance will be lost.[95] This is the price of climate change. From another perspective, these people and practices may survive in a different form that benefits from societal resilience and cultural transformation. Scholars have noted that societies throughout time have "responded to environmental crises [through] innovation [that was] . . . decentralized, protracted, flexible, and broadly based."[96] They have noted that contemporary public discourse often fails to "appreciate that resilience and readaptation depend on identified options, improved understanding, cultural solidarity, enlightened leadership, opportunities for participation and fresh ideas."[97] This book seeks to optimistically highlight a few fresh ideas that can facilitate the adaptations that shape how and where people will live.

2
Climate Change and the Built Environment

The built environment encompasses everything that people call home. It is the physical and material manifestation of where climigrants are leaving from (*sending zones*) and where they are going to (*receiving zones*). The built environment covers everything from buildings and infrastructure to the managed landscapes that support human habitation, commerce, socialization, and leisure.[1] The built environment plays an important role in both causing and adapting to climate change. This chapter looks at the physical and material footprint of the built environment and the role that it plays in driving GHG emissions. This chapter also explores how climate change impacts the built environment. Understanding the complex nature of these impacts helps frame the widespread vulnerabilities of people and firms to a changing environment. It is these direct and indirect impacts that are increasingly playing an important role (e.g., push factor) in shaping decisions to relocate or migrate.

The Built Environment's Role in Driving Climate Change

Some anthropologists suggest that the built environment is a mere artifact of culture,[2] but the reality is that the built environment is home not just to people but a wide range of species, organisms, and urban ecologies that are inseparable from natural ecologies.[3] As such, the footprint of the constructed built environment includes a range of impacts, including impacts that are accelerating the biodiversity crisis facing the planet.[4] Within this urban ecology, it is not just infrastructure services that sustain people. Ecosystem services are so fundamental to human existence that their conservative economic value is equal to the global gross domestic product (GDP).[5]

North. Jesse M. Keenan, Oxford University Press. © Oxford University Press (2025).
DOI: 10.1093/9780197641644.003.0003

While just 1% of the earth's ice-free surface area is covered by infrastructure,[6] 40 million miles of roads have given people access to 90% of the earth's terrestrial surface.[7] At the dawn of the Anthropocene, a vast majority of the earth's land area is managed by people, and the boundaries between the built and natural environments are often functionally indistinct. For most people, these distinctions are largely shaped by their own psychological[8] and aesthetic conceptualization[9] of what the natural environment should look like. The built environment is nearly omnipresent in most people's lives—even when they are in the middle of America's great National Parks.[10]

The global production and use of the built environment accounts for approximately 40% of material consumption, 40% of waste, and upward of 76% of all global energy demand.[11] With more than 147 million housing units,[12]American residential demand is just under 20% of total U.S. energy consumption,[13] while 5.9 million nonresidential buildings with 8.9 billion m^2 (96 billion ft^2) of space account for 17% of all U.S. energy consumption.[14] Manufacturing is attributable to 12% of total U.S. GDP (2021),[15] yet manufacturing facilities consume just 2% of total domestic energy demand. Rather, it is the high-heat processes of chemical and petroleum refining; iron, steel, and cement production; and food and beverage processing that account for half of the primary energy demand associated with the industrial sector's overall consumption of 36% of domestic energy consumption.[16] All of this economic activity requires the built environment.

When one adds the 37% of energy consumption for transportation to the mix, one can argue that nearly all of the energy that is produced and consumed flows through the built environment at some point, including the 421 trillion calories that Americans consume every year.[17] That is equivalent to a diet of about 2,269 Big Macs' worth of calories a year for every individual in the United States. Aside from food, Americans grossly over-consume energy at about 295 gigajoules (GJ) per capita per year,[18] almost twice the per capita energy consumption of Japan and almost 12 times that of India.[19] While this rate is declining, Americans are still responsible for 13 metric tons of carbon dioxide equivalent (MTCO2e) emissions per capita, which is 3 times the global average.[20]

Overall, the built environment formally accounts for about a third of GHG emissions,[21] with the direct construction and indirect operations of buildings accounting for about 21% of total global GHG emissions.[22] The allocation of emissions from urban areas is tricky, as the "precise proportion of emissions from urban areas depends on their definition[,] as well as the attribution of emissions from consumption (upstream), waste (downstream), and the import and export of goods and services (indirect emissions) to urban areas."[23] By one measure in the United States, the "largest 10 cities plus the top 5% of suburbs [account] for more than half of all emissions in the country."[24]

Today, approximately 80% of the U.S. population lives in an urban area.[25] This a far cry from the 5% of the population that lived in urban areas in 1790,[26] where the anti-urban principles of many of the founding fathers came to fruition after observing the political instability of a rapidly urbanizing Europe.[27] Almost 60% of the world's population lives in urban environments, and these areas require vast amounts of land and infrastructure for energy, food, and waste that extend far beyond their immediate borders.[28] The metropolitan region of New York City needs more than 5,000 km^2 (1,930 sq mi) of watershed land for drinking water[29] and the land area equivalent to the entire state of Nebraska just to meet its demand for bread and pasta.[30] Urban areas often require land and infrastructure whose total footprint is frequently hundreds of times the size of the supported jurisdictions.[31]

If business-as-usual urbanization is a planetary pathology, then concrete is cancer. In the next few decades, the total mass of concrete on earth is projected to exceed the total mass of all natural biomass.[32] Concrete and cement alone account for approximately 7% of GHG emissions.[33] Every newborn in the United States will consume more than 25,156 kg (55,461 lbs) of concrete over their lifetime.[34] That is almost two mixer trucks' worth of concrete. But this figure pales compared to the 616,885 kg (1,360,000 lbs) of stone, sand, and gravel that these children will grow up and consume. That is roughly equivalent to almost fifty fully loaded dump trucks.

The built environment requires an enormous amount of material that has to be processed, assembled, shipped, constructed, and disposed of. In the United States alone, construction and demolition waste (C&DW) produces about 544 million mt (600 million tn) a year.[35] That is roughly equivalent to a year's worth of trash in more than 700 large landfills. By one rough measure,[36] this approximately equal to 8.5 MMTCO2e emissions, or the fuel burned by over 2 million passenger cars in a single year.[37] Although about 25% of all C&DW—mostly concrete and rebar—is recycled, the rest of this waste material makes up about 35% of all landfill waste.[38] Intervening hurricanes can quickly multiple these figures, and, as has often been the case in recent history, much of this debris is illegally burned.

In the context of sending zones, abandoned areas represent the prospective waste of huge amounts of material. Although there have been advances in recycling and material technologies,[39] there simply might not be enough landfills to responsibly process all of the building waste from climate impacts.[40] One study published by the real estate firm Zillow estimated that 80,000 homes in Louisiana would be lost with 1.8m (6 ft) of SLR.[41] Assuming that most of these houses are on a concrete slab and weighting for multi-family buildings and average building sizes in the state, one can assume 561 kg per m^2 (115 lbs per ft^2) of CD&W.[42] The resulting loss of an estimated 13.3 million m^2 (144 million ft^2) of housing would create 7.4 billion kg (8.2 million tn) of waste. That is well over a half a million dump truck loads of material waste.

The total resource flows from deconstruction are likely to be a small percentage of the total material going into new construction as people relocate, but it will require coordinated building, engineering, and material codes to fully take advantage of recyclable materials.[43] As construction resources have become more scarce, the economic case for recycling has been observed to be increasingly profitable.[44] Otherwise, for buildings, infrastructure, and even landfills left behind,[45] this material is likely to serve as a significant source of toxic contamination, particularly water-soluble toxics diffused by flooding and SLR inundation.[46] This represents a particularly serious situation for people who lack the capacity to move and are increasingly finding themselves in high-risk sending zones with the growing risk of toxic contamination.[47]

Even today, one can think of a variety of beaches in Texas, Florida, North Carolina, and Massachusetts that are littered with dangerous debris from homes that have fallen into the ocean. During hurricane season, it is not uncommon for beachgoers along Florida's coasts to find entire assemblies of homes floating ashore. As toxic pollution and hazardous debris converge along America's coasts, the amenity value of beaches to local tourism economies are likely to wane.[48] This phenomenon represents a potential feedback loop where declining tax revenue from tourism combines with declining property tax revenue to limit the local capacity for scaled adaptation and environmental cleanup interventions. In addition, the impairment of local coastal ecosystems, including mangroves and marshes, threatens to reverse valuable carbon sinks.[49]

In the United States, much of the population growth this century has been concentrated in the Sun Belt. These geographies are uniquely defined by a range of climate risks from droughts in the Southwest to more intense hurricanes in the Southeast. But these places also represent a unique carbon and GHG story. Their rapid growth since the 1990s has been defined by increasingly large and energy-intensive homes in car-dependent suburbs.[50] Residents in these parts of the country spend more time in their car commuting,[51] and they rely significantly less on mass transit.[52] Their sources of power are often more likely to come from coal and natural gas plants.[53] Although the overall energy intensity of the housing in these regions is lower than the Midwest and Northeast because they have benefited from newer housing with updated energy codes, their overall GHG intensity is comparatively high because of the convergence of dirty power, sprawling land uses, and car-centered transportation.[54]

These externalities are part of the equation that made Sun Belt housing originally so cheap. As cited in chapter 1, these externalities are increasingly being internalized in a manner that is working to increase costs. For many, it is hard to imagine that any climate impact could reverse the growth of the Atlanta suburbs—unless, of course, they run out of water.[55] Overall, suburbs have a much higher carbon footprint, with up to four times the GHG emissions of nearby urban cores.[56] While cities often have old, leaky energy-inefficient housing, the research is clear—greater population density drives down per capita carbon emissions.[57] As people relocate, the question is whether they will bring their carbon-intensive lifestyles and models of the built environment with them to receiving zones. As will be explored in chapter 6, there are models for sustainable design and development that offer something profoundly more responsible for receiving zones. The goal is not just to adapt to climate impacts but to mitigate the further acceleration of climate change by structurally decarbonizing the built environment.

Climate Vulnerability and Exposure in the Built Environment

The built environment faces a variety of exposures to climate change impacts—both in terms of the shocks of extreme events and the chronic stresses that come along with a changing operational environment. As housing and infrastructure services begin to degrade and fail, people and firms have several fundamental options, including the ultimate option to relocate. They can invest in the capacity of infrastructure systems to adapt and maintain engineering resilience. But there are limits to the resources necessary to make these investments, and there are thresholds from which systems are no longer functional or economically viable.

As is often the case, people, governments, and utilities make just enough investments in adaptation and hazard risk reduction to buy time and delay the hard decisions. The politics of climate change and the built environment are local and full of conflicts. As one mayor in South Florida once said, "The true success of local climate leadership is letting people down in increments that they can absorb."[58]

Understanding the nature of exposure and vulnerability in the built environment is critical for informing and guiding various adaptation policies and plans that are explored in more detail in chapter 3. Climate impacts are not just impairing and destroying infrastructure, they are also changing the provision of critical environmental resources—like drinking water—that often provided the very logic for the foundation of urban centers. The good news is that the state of adaptation practice in planning, design, and engineering is rapidly progressing. The less good news is that the direct and indirect impacts from climate change are sometimes difficult to project with the spatial and temporal precision necessary to efficiently and effectively allocate limited public resources.

Whether people realize it or not, they are already paying for these emerging adaptation, resilience, and recovery costs in their utility and tax bills, and this includes everything from higher utility rates to the credit burdens associated with utility disconnections for cost-burdened households.[59] As will be explored in chapter 4, they are also paying for these costs in the private market for everything from housing to insurance. For infrastructure providers, the financial costs of extreme events, like floods and hurricanes, can be financially debilitating to the extent that it limits their capacity to invest in the long term and maintain their assets in the near term. While many local governments and utilities are slow to account for climate change, the market is increasingly sending economic signals that cannot be overlooked.

Since 2015, many well-resourced jurisdictions have invested in the production of long-term climate vulnerability assessments. These studies have evaluated a wide variety of structures and infrastructure systems—from energy grids to drinking water systems. Some of these jurisdictions have even developed adaptation and resilience plans.[60] Unfortunately, as will be outlined in chapter 3, a vast majority of local jurisdictions are woefully behind on planning for and investing in climate adaptation.[61] Most jurisdictions are flying blind as to the localized risks of climate change in the built environment, and this could very well accelerate climigration and relocation in the future. In the near term, climate impacts are already a powerful push factor for many households.

Extreme Precipitation, Flooding, and Moisture

The challenge with climate change is that there is either too much water or not enough. The average return intervals (ARI) of extreme precipitation events in the United States have increased in their frequency and intensity,[62] and they are projected to increase in the future.[63] As the atmosphere heats up, it can hold more moisture. For every one degree Celsius that the planet's atmospheric temperature rises, the amount of water vapor in the atmosphere can increase by approximately 7%.[64] The atmosphere draws that moisture up from everything from the oceans to plants and soil. In some places, this increases the risk of drought. In other places, when it rains, it pours. Map 2.1 highlights how much more frequent extreme precipitation events are likely to be across the continental United States. Approximately three-quarters all of disaster declarations in the United States stem from flooding.[65] Flooding is the second leading cause of weather-related deaths, with around a 100 people drowning across the country every year[66]—often in their cars.[67]

In June 2024, places in South Florida received just over .3 m (2 ft) of rain over the course of several days.[68] At times, the rain was coming down at a rate of 10 cm (4 in) to 15 cm (6 in) an hour. The ARI for this kind of storm is about 1 in 500 years. The bigger problem is that Florida has had one 500-year event every year over the past four years (2020–2024). This is by no means a fluke event. In 2017, Hurricane Harvey hit the Texas coast and dropped 153.87 cm (60.58 in) of rain over a five-day period in Nederland, just north of Houston.[69] Subsequent climate-attribution research found that "climate change likely increased the chances of the observed rainfall by a factor of at least 3.5."[70]

Observed precipitation trends over the thirty-year period of 1991–2020 suggest that the eastern part of the country is getting wetter and the western part of the country is getting slightly drier.[71] As the *Fifth National Climate Assessment* (*NCA5*) noted, "Average annual precipitation from 2002–2021

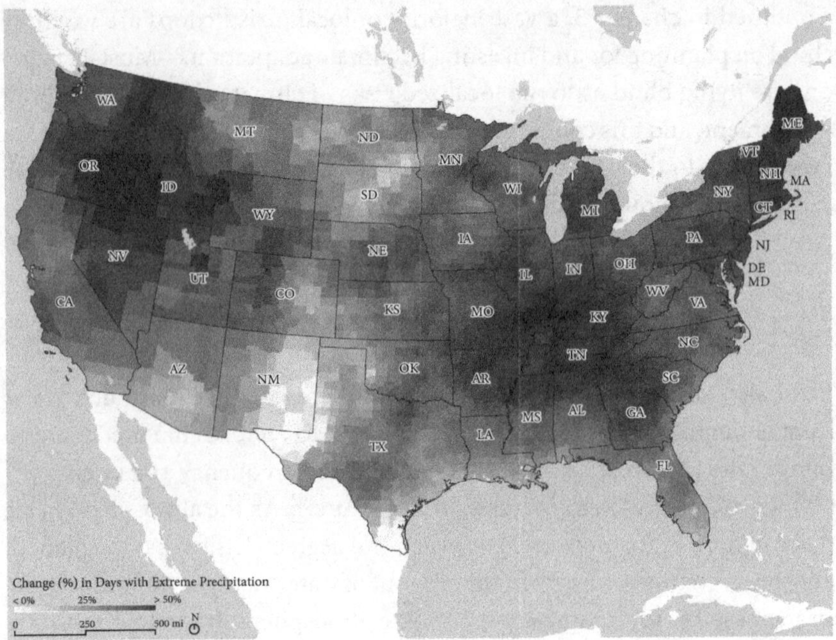

Map 2.1 Projected Change (%) in Days with Extreme Precipitation from 1991–2021 Baseline with 3°C (5.4°F) of Global Warming. Design Credit: Oliver Oglesby and Jesse M. Keenan. Data Source: U.S. Global Change Research Program. 2023. National Climate Assessment: Interactive Atlas Explorer. Washington, D.C.: U.S. Global Change Research Program.

was 5%–15% higher relative to the 1901–1960 average in the central and eastern U.S., a trend attributable to climate change."[72] Looking forward, Map 2.2 highlights the overall projected change in annual precipitation. But this is a reflection of total precipitation and does not speak to the timing of this increased precipitation. In a stark warning, the *NCA5* stated that "extreme precipitation-producing weather systems ranging from tropical cyclones to atmospheric rivers are *very likely* to produce heavier precipitation."[73] When it comes to extremes, the records continue to be broken. In 2018, Kauai, Hawaii, broke the record for the most rain in a 24-hour period with 126.21 cm (49.69 in).[74]

It is not just extreme rain events. Climate change is also altering the type and ARI of different types of precipitation. Although hail is notoriously difficult to model, it is "generally anticipated that low-level moisture and convective instability" will increase the likelihood of hail events and result in potentially larger hailstones.[75] In 2023, the insurance industry reported

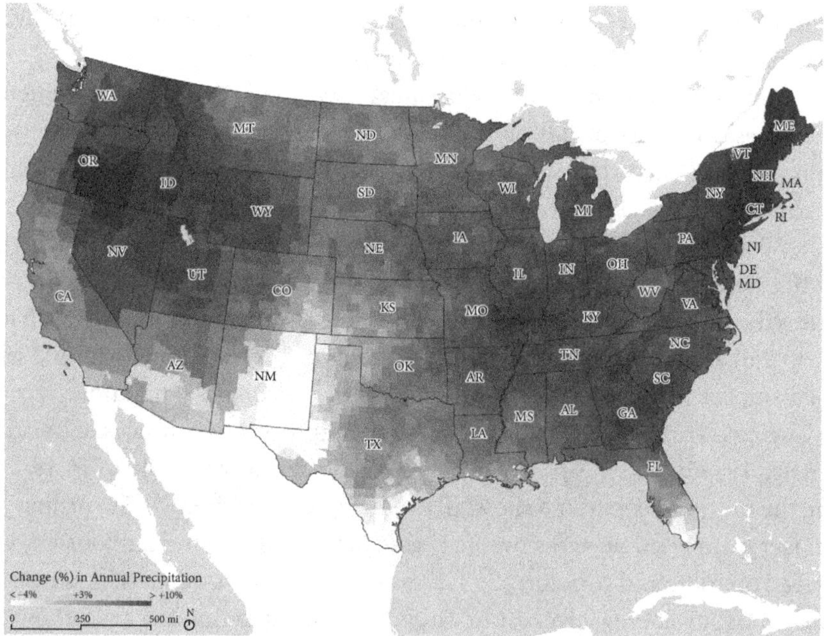

Map 2.2 Projected Change (%) in Annual Precipitation from 1991–2021 Baseline with 3°C (5.4°F) of Global Warming. Design Credit: Oliver Oglesby and Jesse M. Keenan. Data Source: U.S. Global Change Research Program. 2023. National Climate Assessment: Interactive Atlas Explorer. Washington, D.C.: U.S. Global Change Research Program.

more than $60 billion in losses from convective storms.[76] Up until the 2010s, the insurance industry averaged about $1 billion in losses every year associated with hail events.[77] Today, the industry averages more than 500,000 claims totaling between $8 and $15 billion every year—accounting for 45% of *all* homeowners insurance claims. A single hail event in Texas and Oklahoma in 2021 caused $3.3 billion in damage.[78] Most of the damages to buildings are to roofs and windows, with damages varying greatly depending on the nature of the roofing material (e.g., shingles, graveled tar, tiled ceramics) and the pitch of the roof.[79]

In some places, climate change is also shifting precipitation from snow to ice.[80] Ice and freezing rain cause damage to telephone and powerlines, grid infrastructure, and trees.[81] The damage to trees from ice and windstorms can undermine natural radiant barriers that mitigate the impacts of the urban heat island effect. To demonstrate how complex the interaction of earth and built environment systems are, the urban heat island effect itself can drive increased precipitation, particularly in the summer.[82]

As ice events become more widespread, transportation managers often use excess amounts of salt on the roads, which may dangerously increase salinity levels in surrounding water bodies.[83] The increased salinity from road de-icing was one of the major drivers of the Flint, Michigan, drinking water crisis: Salty river water was improperly treated, and the resulting pH of the water caused lead leaching.[84] Flint and other domestic urban water crises demonstrate the true fragility of the critical infrastructure systems that people depend on. As drinking water systems begin to be compromised in high-impact sending zones, the clock starts ticking on out-migration.

The cold truth of climate impacts is seemingly full of contrasts. Climate change is reducing mean snowfall,[85] but it is also attributable to increasing the number of extreme snowfall events in some regions.[86] The primary impact from extreme snow events is associated with the tremendous loads placed on roofs and structures.[87] The groundbreaking research of the Arctic Design Group at the University of Virginia School of Architecture has shown that increased snowfall in Alaska acts as an insulator from the super-cold ambient air temperatures and prevents the ground from fully refreezing in the winter. Refreezing is critical because the permafrost is thawing at greater rates and depths in the summer, compromising the structural integrity of foundations supporting housing and critical infrastructure, including national security infrastructure.[88]

Even seemingly benign phenomena such as an increase in late spring frosts attributable to climate change[89] can lead to serious damage to buildings, including damage to porous exterior materials associated with many types of masonry and concrete.[90] Damage to these materials can, in turn, undermine the thermal performance of these buildings, which may increase energy costs for heating and cooling. It may also undermine the capacity of these buildings to maintain a habitable temperature when the power goes out.

ARI is an important metric in the engineering and design of the built environment. In the context of water, it speaks to the upper boundaries of potential wastewater, stormwater, and even meltwater volume that need to be accounted for in the sizing of everything from culverts to gutters. Buildings need wider gutters to handle more rain. When there is more rain than the gutters can absorb, then the water backs up and can cause damage to the building. The same problem happens on an infrastructural scale.

In a best-case scenario, local governments incorporate local and regional climate projections into their "design event" requirements. For instance, in New York City (NYC) a building project might have to accommodate the stormwater from a 24-hour duration rainfall event with a 100-year ARI. By today's standard that might be about 20.3 cm (8 in), but by the end of this century that number goes up to more than 33.0 cm (13 in).[91] Unfortunately, most jurisdictions do not have the scientific and engineering resources that NYC has. As has been observed, "Considerable mismatch exists between climate model outputs and the data inputs needed for engineering designs."[92] Even when engineering standards incorporate statistical properties for future time periods, different interrelated infrastructure systems may have fundamentally different approaches. For instance, the exceedance probabilities in transportation design differ significantly from most stormwater designs even though transportation infrastructure often serves a dual function of steering, slowing, and storing stormwater.[93] Translating cutting-edge climate science into laws, regulations, and engineering guidelines takes people, money, and political will, and a vast majority of jurisdictions are largely flying blind.

While changing precipitation patterns represent a substantial range of impacts, the most worrisome impact is the risk of flooding. It is not just the frequency of extreme precipitation events but the novelty of where the flooding is happening. Flood maps in the United States vary from old paper maps from the 1970s to high-quality state-of-the-art LIDAR- and simulation-informed maps. Many jurisdictions simply do not have the money or resources to adequately map their flood risk. The single greatest source of flood map information comes from the National Flood Insurance Program (NFIP), and these mapping exercises (Flood Insurance Rate Maps [FIRMs]) are ultimately political negotiations.[94]

The Federal Emergency Management Agency (FEMA), as administrator of the NFIP, frequently tried—prior to the second Trump administration—its best to make the maps more accurate and to update them to account for climate change. To counter these efforts, local governments often went through an administrative process to appeal and remove locations from the FIRMs. Local governments did this because local politicians bear the political consequence of new flood insurance requirements that come along with being in a flood zone. Even when NYC endeavored to expand the flood maps to include the latest SLR projections after Hurricane Sandy, they later regretted this initiative and took actions to appeal and reduce the area included in the expanded flood zones—even though they knew these areas are likely to be flooded and/or inundated in the future.[95]

With or without NFIP-driven flood mapping, projecting future flooding with climate change is a highly active area of science. Flooding can manifest in a lot of different ways aside from rivers overtopping their banks and storm surge. Flooding can also happen from the upward movement of the water table and via the conduits of stormwater infrastructure itself. The impacts of flooding on the built environment vary a good bit. Sometimes minor flooding can scour a building's or bridge's foundations and wash away structural soils that weaken the structure.[96] In other cases, the buoyancy effect of water can literally lift structures out of the ground.[97] If the floodwaters contain any portion of saltwater, then the corrosive effects of salt on the rebar and concrete can lead to long-term structural damage. Although initially designed to preserve limited supplies of freshwater sand, new technologies in cement allow for the use of seawater in the admixture designed to withstand these corrosive effects.[98] Floodwaters also carry sediments and debris that can damage buildings, bridges, roads, dams, and other infrastructure.

Dams are particularly vulnerability to both the high and low levels of water that come with greater variability in precipitation.[99] While there are a number of high-profile large public dams at risk, much of the risk from dam failure is centered on the hundreds and often thousands of private dams that exist in any given state—many of which are not maintained or inspected.[100] To make matters more challenging, American dams are getting relatively old, with the average age being around 60.[101]

In Alaska, melting ice from glaciers often coincides with spring snowmelt to create a natural dam.[102] When these dams explode, one does not want to be anywhere near the shrapnel of ice and biomass. The good news is that Florida does not have to worry about ice dams. The bad news is that the State of Florida does not actually know how many dams are in the state. It has records for more than 1,000 dams, but the state Department of Environmental Protection speculates that this number is not accurate.[103] Initially developed to manage potential flooding, some quantitative research in Florida has found that "dams do not significantly reduce flood damage."[104] Whether it is a dam failure or just a flash flood, if anyone has ever seen someone's home float at high-speed into a bridge, they will never forget it.

There are also indirect impacts from the flooding of infrastructure, such as when compromised pumping, power, and communications equipment causes dependent infrastructure to go offline, overheat, and/or malfunction. From a system-of-systems (SoS) point of view, these impairments and failures can extend to everything from energy to healthcare systems. Hospitals are a perfect example of a contextual SoS vulnerability given their wide-ranging activities from patient care to commercial biomedical research. While engineering resilience in healthcare facilities has significantly improved following flood events at major hospitals in Houston and NYC,[105] there is still an increasing likelihood that people are going to die and valuable research is going to be lost in future flood events.[106] In Florida, 12% of hospitals are directly in flood zones.[107] This includes the largest medical centers in both Miami (Jackson Memorial Hospital) and Tampa (Tampa General Hospital) that are not only located in flood zones; they are also located in SLR zones. You can take a boat to both facilities.

The acute and chronic health impacts from flooding and climate change can be serious. People drown, they slip and break bones, and they cut themselves in the cleanup process. In sending zones with an older average age, there may be a disproportionate impact on the population. Although seniors may have more lifelong coping experience and accumulated wealth, their capacity to physically endure flood events may be impaired, particularly when it comes to the cleanup and rebuilding.[108] Just filing the insurance claim paperwork can be extremely stressful.

It is not just seniors who are at risk. Although less than 10% of the U.S. population directly lives in a flood zone, this population is disproportionally composed of economically disadvantaged people[109] that are already bearing the weight of limited healthcare access and increased healthcare costs.[110] As healthcare becomes more difficult to access in sending zones, some people, particularly those with chronic health conditions, may have no other choice but to leave. Interviews on Cape Cod, Massachusetts, noted that high housing and insurance costs—partially attributable to flooding and storm impacts—were driving healthcare providers off the Cape. Interviewees feared that this exodus of providers would hurt the economic attractiveness of the region and raise overall costs. In this light, can Florida maintain its mental monopoly for fixed-income retirees and working-class people? These kinds of economic stresses certainly do not help. Worse, economically disadvantaged households can least afford the time, much less the cost, that it takes to manage chronic health conditions, particularly those conditions that may arise from floods and increased moisture in the built environment (e.g., mold-driven asthma).

Flood waters may contain a wide variety of biological and toxic contaminants, often through the spread of untreated sewer water, human waste, and dead animals. Buildings have a surprisingly complex and diverse microbiome[111] where occupants and buildings regularly transfer microbes.[112] But this kind of microbial contamination represents a substantive hazard for occupants that may linger well after the flood.[113] The most challenging long-term biological hazard from floods is mold. Not all mold is dangerous, and some species may even be beneficial for building occupants.[114] While every building will have some form of mold, the mold found in flooded buildings often contains species with mycotoxins that impact human health.[115] Usually, the higher the flood, the greater the amount of mold in a building.[116]

Beyond flooding, the increased levels of moisture in the atmosphere with climate change also pose a challenge because they can stimulate mold growth.[117] Particularly in historically less humid and cooler latitudes, building codes may not require vapor barriers that prevent moisture from entering a building's interior. Moisture may penetrate the building from direct contact on the surface of the exterior, such as when gutters fail. More often, moisture penetrates by virtue of the pressure differential between the inside and outside of the building, wherein moist air is drawn into the low pressure (created by mechanical ventilation) of the interior of the building.

A combination of outdated building and mechanical codes and inferior construction means that the risk of mold in housing is moving farther north.[118] In 2021, the American Society of Heating, Refrigerating and Air-Conditioning Engineers (ASHRAE) updated their building code reference maps to account for, among other things, an increasingly warmer climate, particularly in the northern parts of the United States.[119] As places such as southern Wisconsin quickly warm, there are new standards for temperature and humidity control.

It is not just places with higher humidity that are at risk from mold. More alarmingly, the Southwest is not immune to the spread of dangerous northern mold. The increased spread of the deadly valley fever respiratory disease (coccidioidomycosis) northward has been attributable to climate change.[120] Although rare, this disease has a mortality rate of upward of 10%–20%.[121]

The more immediate health challenge associated with mold is asthma. Not only is asthma a devastating challenge for children, who are often exposed to mold in school buildings,[122] but claims and litigation from mold represent a major economic drain on insurers and landlords.[123] It is not uncommon for mold settlements to reach millions of dollars. This sometimes provides incentives for landlords and property owners to mitigate growth and exposure.[124] To facilitate this effort, there is a new wave of pro-tenant mold regulation being implemented within state and local jurisdictions.[125] These regulatory efforts include transactional mold disclosure for renters and purchasers, specific measurement standards for mold and dampness, and greater enforcement power for local building code officials. In some jurisdictions, such as Washington, D.C., a failure to properly manage mold may result in costly penalties.[126] Unfortunately, Washington, D.C., has a bigger problem—the swamp is sinking, and the water is rising.

Sea Level Rise, Inundation, and Tropical Cyclones

It is hard to fathom the scale of SLR, yet SLR is often the starting point for examining the demographic implications of climigration.[127] This is not a novel area of scientific inquiry. Paleoclimatology and archaeology have provided a wide range of evidence to understand human migration responses to SLR from the Pleistocene and Early Holocene forward.[128] Even at the time of the earliest human migrations to North America 20,000 B.P., SLR was shaping the land and coastal routes taken by early settlers.[129]

Starting around 4,500 B.P., there is robust archaeological evidence of coastal adaptations among native populations in Florida, including the siting of settlements, the exploitation of shellfish harvesting, and the construction of earthen structures likely to manage flooding.[130] During major fluctuations in sea levels between 3,400 and 2,400 B.P., particularly along the Gulf Coast, evidence suggested that native populations migrated inland, shifted their technologies toward hunting and riverine fishing, and stopped building large monumental constructions on the coast.[131] Although there are unresolved debates on the sedentary nature of some native populations—thanks to pioneering work of archaeologist Victor Thompson and others—there is a robust archaeological record that provides evidence of the earliest known climigration in Florida. Even as people were merely moving off the immediate coasts, their built environment was adapting to new technologies, materials, and environmental interactions. Across the board, paleoclimatic research has offered a stark reflection on a range of devastating climate impacts that even a moderate amount of warming can bring.[132]

In contemporary terms, part of the reason that SLR is attractive to researchers and policy makers is that inundation from SLR is seemingly so definitive and conceptually clear. If someone is living squarely in an SLR zone, they will not be living there in the future. Even if they lived on a boat, submerged land rights are quite distinct from land rights.[133] Yet the timing of SLR inundation is not always so clear. One problem is that economics tells us that there is a difference between absolute inundation and effective inundation, with one measure of effective inundation being properties that are at least partially flooded upward of 26 times a year.[134] In this sense, the built environment's vulnerability to SLR should be understood along a continuum of adaptive and maladaptive behaviors ranging from orderly, managed relocation to inhabited soggy, unsanitary occupation on the fringes of effective inundation. More than one coastal SLR scenario involves the inhabitation of immobile people in the landscapes of post-relocation sending zones of the wealthy.

The built environment's exposure to SLR in the United States was first comprehensively studied by the U.S. Environmental Protection Agency (EPA) in the early 1980s.[135] With quantitative tools rudimentary by today's standards, the EPA's work was remarkable for its methodological clarity and foresight. The authors noted:

> Many decisions have outcomes that last long enough to be affected by [SLR, including] where to locate roads, wastewater treatment plans, and chemical and nuclear waste facilities. A coastal highway may determine development patterns long after the pavement and the structures along the road have been replaced. . . . [P]rudence demands that decision makers plan for at least the low [SLR] scenario.[136]

SLR—or technically mean relative sea level rise (MRSLR)—is a calculation of the height of sea levels relative to the vertical upward or downward movement of coastal land. In many places where the land is sinking, such as in southeast Louisiana, the net effect is a faster rate of MRSLR than in places where the land is vertically stable. New Orleans is in a particularly difficult situation because the weight of its post-Katrina levee system is so great that it is causing the structures to sink just as the seas are starting to rise. By the assessment of U.S. Army Corps of Engineers (USACE) itself, the system is likely already statistically inadequate, and it will need to be continually elevated and reinforced, at a huge expense, for an indefinite amount of time.[137] In some places such as Alaska, the land is actually rebounding and moving vertically upward after thousands of years of being compressed by glaciers. In these select locations, relative sea levels may actually be going down. Unfortunately, in many places from the Mid-Atlantic to the Texas Gulf Coast, the land is sinking.

Global mean SLR has increased over the decades, with a current annual observed rate of increase, in the era of satellites, of 3.4 mm ± .4 mm (.13 in ± .01 in) from 1993 to 2022.[138] Just from 2013 to 2022, the global annual rate of SLR came in at around 4.4 mm (.17 in) per year.[139] In the United States, MRSLR has varied greatly for a variety of reasons, including variable offshore currents and land subsidence from natural processes, groundwater withdrawal, and fossil fuel extraction.[140]

Louisiana is a perfect case study for the confluence of bad luck and bad decisions. Oil and gas drilling destroyed vast amounts of coastal wetlands, and this has contributed to land loss—some 5,100 km^2 (2,000 sq mi) since 1930s.[141] This land loss led to creeping salinity that choked off sensitive ecologies and further accelerated land loss processes. One adds natural land processes, human development, and the natural variability in the relatively shallow Gulf of Mexico, and the rates of MRSLR are around 12.7 mm (.5 in) a year.[142] From North Carolina to Texas, the rates of MRSLR far exceed global and national rates.

It is worth noting that the water levels of the Great Lakes are also projected to rise from increased precipitation over the lakes and around the lakes' watersheds.[143] In the case of Lake Superior, water levels on the southern shore are anticipated to increase, in part, due to a gradual titling (i.e., isostatic rebounding from glaciers) to the south of the underlying geology.[144] As highlighted in chapter 6, this is precisely where Duluth, Minnesota, is located. Even with approximately 10% of the world's surface freshwater, there might be a little too much liquid gold on the horizon for newly arrived climigrants. As lake levels are regulated by the federal government, this lake level rise will have to be accounted for, particularly in the management of hydroelectric power.[145] Unfortunately, the federal government does not have a mechanism for regulating sea levels—other than perhaps wholesale decarbonization.

The future rate of SLR is dependent on a lot of factors. Aside from relative factors like subsidence, other factors include natural variability, tidal processes, ice mass loss, GHG emissions, and the amount of energy and heat the oceans can absorb.[146] Up until now, the thermal expansion of the water in the world's oceans accounted for about half of SLR, and meltwater from ice and glaciers makes up most of the difference. Upward of 91% of the excess heat energy from climate change has been absorbed by the oceans, and there are some fundamental questions about how much more heat that the oceans can absorb from the atmosphere before crossing a critical threshold.[147] In one scenario, that threshold might mean the weakening or collapse of ocean currents that are critical for mediating current weather patterns.

Projecting SLR in the United States is a complicated business, and different academic perspectives and federal agency points of view have resulted in a variety of projections. This is not necessarily a bad thing, as it allows for robust comparisons of models and results. However, state and local officials are increasingly looking for clarity and direction on the levels of SLR they should plan for. Although NYC, Boston, Tampa, and counties in the Southeast Florida Regional Climate Compact around Miami have undertaken downscaled regional projections to guide local planning, most jurisdictions do not have these kinds of resources.

To address this challenge, the Biden administration created the U.S. Interagency Sea Level Rise Task Force. The task force produced five projections for SLR up to the year 2100. The low projection is for .3 m (1 ft) and the high projection is for 2.0 m (6.6 ft).[148] This high projection is not a worst-case scenario that might be associated with a rapid ice melt scenario, such as rapid melting in the West Antarctic ice-shelf.[149] While scientific consensus varies, engineering practice in NYC generally assigns around a 10% probability to expected value calculations associated with protective infrastructure design to account for a rapid ice melt scenario.[150] This is based more on the precautionary principle than the science.

In the context of climate change, a shift in the mean usually corresponds with a flattening of the statistical distribution. With a focus on average shifts in things like SLR, society overlooks the tail risks that represent an increased chance for extreme phenomena, such as a rapid ice melt scenario. By the same token, the year 2100 has been a convenient cutoff year for climate projections, yet most of the buildings and infrastructure being planned today will have a useful life well beyond this point. This reinforces the proposition that it is not just how one designs a building; it is also about where one sites that building. The flip side is that the year 2100 has been a convenient benchmark that is within the realm of people's capacity to look forward in time. Climate policy has a lot to do with communications, especially when people have to internalize the ideas of slow violence. Human brains evolved for fast violence—not slow violence.[151]

Over time, SLR gradually interacts with a variety of climatic and meteorological conditions. For instance, nuisance floods have become more severe and more frequent. What was once a 1-in-500-year ARI might now be a 1-in-50-year ARI. When storm surges do hit, they are riding on top of increased sea levels that amplify the intensity of the force and expand the impacted geography. The overall increase in sea levels may limit the capacity of stormwater to discharge, as pipes simply get inundated with or without backflow or check values that force water to flow in one direction.

SLR also pushes saltwater into groundwater, which leads to a number of contamination challenges. Hydrostatic forces themselves can shift soil and significantly damage infrastructure. Water with varying levels of salinity can penetrate buildings through their foundations or it can be blown onto exteriors. Salt can slowly creep into masonry and through cycles of evaporation damage building materials and structural elements.[152] Salts increase the pH of materials, which can accelerate degradation and aging effects. Salt exposure quite simply causes the chemical bonds to break down, particularly in concrete. Increased salinity also increases the corrosive effects on pipes, as "there is a direct correlation between soil corrosiveness and break rates of metallic pipes."[153] Sometimes SLR can even drive leaks of sewage directly into public spaces.[154]

Broken drinking water pipes have wide-ranging implications from increased risks of biological contamination to significantly increased energy demands required for processing and pumping more water. More than half of the drinking water processed in New Orleans is lost in a vast system of broken pipes and leaks. This is referred to in the industry as "non-revenue water." Lab and home testing of total chlorine residuals in New Orleans often highlights unhealthy elevated levels of chemicals used to guard against biological contamination.[155] Of course, when one is looking to kill the brain-eating amoeba *Naegleria fowleri*, slowly impairing the health of residents with excess decontaminants might be a reasonable trade-off.

The relative difference in density between fresh and salt water can also push groundwater upward in a manner that exacerbates flooding and compromises drinking water.[156] Flooding is amplified because a high water table limits the total amount of rainwater and stormwater that can infiltrate and percolate into the soil or flow into neighboring water bodies. A combination of increased precipitation, higher water tables, and SLR also represents an increased risk of the migration of toxic chemicals from waste sites into the drinking water.[157] This means that local water utilities potentially have to move well sites and expand protection zones to account for more-polluted surface water. Miami-Dade County, Florida, has a particularly perverse situation where quarries that are mined to produce fill for elevating the region out of the floodwaters have become conduits for dangerous chemicals seeping into the drinking water.[158] Many people in high-risk areas have to think twice about giving their kids unfiltered water to drink.

In coastal geographies, high water tables can also result in septic tanks leaking waste material and excess nutrients directly into aquifers used for drinking water.[159] This increases the risk of the spread of gastrointestinal diseases and can drive algae blooms that choke off oxygen to fish and marine ecologies. There are millions of septic tanks in the United States, with upward of 20% of households using septic tanks.[160] Miami-Dade County, Florida has more than 120,000 septic tanks at risk from contamination, and many low-income residents often have limited use of their plumbing, particularly after it rains.[161] Connecting to a sewer often costs tens of thousands of dollars per household, and many people simply cannot afford the expense.

For some people, living in Miami means going to the bathroom in a bucket. While extremely uncommon, local cases of cholera (*Vibrio cholerae*) do appear in Florida.[162] A warmer and wetter world significantly raises the risk of waterborne and tropical diseases spreading into the United States. In the Southeast, the emergence of new and old vector-borne diseases is extremely troubling.[163] It is not uncommon for local cases of dengue fever (*Orthoflavivirus dengue*) and malaria (*Plasmodium vivax*) to pop up in Florida, and the environmental conditions are projected to be ripe for a broader endemic spread for these and other diseases in the future.[164] By one estimate, "over half of all human pathogenic diseases can be aggravated by climate change."[165] In the future, people who live in areas with uncontrolled endemic tropical diseases may think twice about everything from going for a walk in the park to having children.

From roads to sea walls, the built environment faces a wide range of risks from SLR. Although historically sited on high elevations, recent generations of road and rail development were often laid out at low elevations along shorelines to save on land procurement costs. When rails flood and become inundated, mass transit riders pay the price. It is often low- to moderate-income riders that bear the highest costs in terms of expense and time.[166] If a train is delayed because of a flood, hourly workers pay for it in lost wages. Many people cannot afford to just take an Uber. It is not just rail and mass transit. Airports are particularly challenging because they are going to need not only massive coastal flood protection systems; they are also going to need to lengthen their runways to accommodate commercial aircraft during periods of extreme heat when there is not sufficient distance to safely accommodate the lift necessary for takeoff.[167] Many of the largest airports in the country are built on top of fill in adjacent water bodies, including airports in Boston; NYC; Philadelphia; Washington, D.C.; and San Francisco. The Louis Armstrong New Orleans International Airport is one of the lowest-lying airports in the world at just 1.4 m (4.5 ft) above sea level. The new billion-dollar terminal continues to sink under its own weight.

Airports, water treatment facilities, and public housing were often sited on fill in low-cost and low-lying coastal areas. A great deal of affordable housing was created in flood and SLR zones because land there is cheap. In real estate, land prices are often the ultimate economic determinant of relative affordability.[168] Almost 10% of all public housing is on a floodplain.[169] Worse, the racial politics of spatial exclusion, in places like Charleston, South Carolina, set up a situation where the most underrepresented classes of people bear the weight of government inaction, negligence, and indifference toward preparing for climate impacts.[170] As Harvard Law's Susan Crawford highlighted, sometimes local governments just look the other way and continue allow building in high-risk areas that are already half underwater because it is someone else's problem.[171] In South Carolina and many other places, poor, rural, and predominately Black communities will disproportionately bear the burden of SLR.[172]

The risk of inundation and the destruction of unique American cultures extends to everything from lost archaeological sites[173] to historically significant buildings, landscapes, and places.[174] The systems of governance associated with historic preservation offer valuable insight into the nature of the vulnerability of the built environment to SLR and flooding. In this light, there is often an intransigence by historic preservation officials to adapt buildings through the utilization of new materials and changes in elevation. In Miami Beach, interviews with preservations suggested that if they loosened up some regulations in the name of climate change, then it would become a slippery slope for developers to gradually unravel the entire systems of historic preservation regulation.[175] For this reason, some scholars are exploring the idea of de-designating historically protected buildings and districts in order to allow for protective adaptation measures that otherwise would have been prohibited due to the measure's lack of historical authenticity. Generations of historic preservation regulations are at risk from unraveling in the name of adaptation.

More than 123 million people, or approximately 40% of the U.S. population, live in coastal counties.[176] Nearly a million Americans are presently exposed to coastal flood events every year, and current projections suggest that upward of 22 million Americans will be living in a coastal floodplain by the end of the century.[177] Early projections suggest that just .9 m (3 ft) of SLR could drive 4 million people to relocate, with 1.8 m (6 ft) of SLR driving upward of 13 million people from their homes.[178] Map 2.3 highlights the vast geography covered by SLR and flood zones, with many high-risk areas in the Southeast overlapping with areas defined by high measures of social vulnerability.

Across this geography, hundreds of thousands of homes are currently at risk of being inundated from SLR alone within their remaining useful life. SLR is not just about the people, homes, and infrastructure that are physically or effectively inundated. When roads, schools, and elements of the built environment that bind societies and economies are compromised, the isolation will disconnect people and place in a manner that is likely to drive significant climigration. This is true even for those living on high ground. Some research suggests that such functional isolation may significantly increase existing projected displaced populations by 30%–90% and will likely drive relocation "decades sooner than the risk of inundation."[179]

Tropical cyclones, hurricanes, and tropical systems have long been a driver of the movement of people in the United States.[180] Sometimes places are permanently wiped off the map. Up until only a few years ago, the drive from Miami to the Florida Keys would pass by rows of palm trees and other decorative plants commercially planted among the remaining concrete foundations in and around Homestead of the thousands of homes left behind by Hurricane Andrew in 1992. A sudden and dramatic rise in elevation—in an otherwise extremely flat landscape—was a stark reminder of the final resting place of the remains of what was the American dream for an awful lot of people. Hundreds of thousands of people left and never came back.[181]

Scientists are still debating the extent to which climate change increases the frequency of tropical cyclones and hurricanes, although some research has modeled favorable atmospheric conditions for developing and steering hurricanes in the Western Atlantic.[182] Much of the focus has been on the increased intensity of destructive hurricanes associated with elevated sea surface temperatures.[183] Over extremely warm waters of the North Atlantic and the Gulf of Mexico, hurricanes are intensifying very rapidly. They are also carrying more moisture. What science can agree on is that "that both climate change ([e.g.,] higher ocean heat, sea surface temperature and cloud cover moisture) and climatic variability ([e.g.,] ENSO and AMO)[184] might

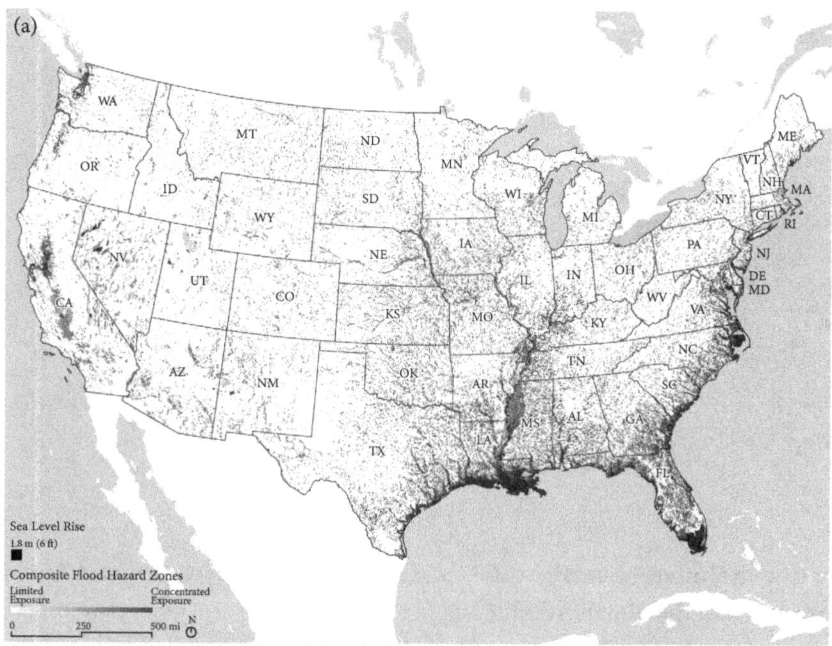

Map 2.3 Flood Risk in the United States: (a) Composite Land Area Exposed to 1.8 m (6 ft) of Sea Level Rise Inundation and Flood Hazard Zones; (b) Convergence of Flood Exposure and Contextual Social Vulnerability. Design Credit: Oliver Oglesby and Jesse M. Keenan. Data Source: (a) National Oceanographic and Atmospheric Administration. 2024. Sea Level Rise Viewer. Charleston, SC: Office of Coastal Management, National Oceanographic and Atmospheric Administration; National Oceanographic and Atmospheric Administration. 2024. Coastal Flood Exposure Mapper. Charleston, SC: Office of Coastal Management, National Oceanographic and Atmospheric Administration; Natural Resource Conservation Service. 2023. Soil Survey Geographic Database (SSURGO). Washington, D.C.: U.S. Department of Agriculture; (b) Tate, Eric, Asif Rahman, Christopher T. Emrich, and Christopher C. Sampson. 2021. "Flood exposure and social vulnerability in the United States." *Natural Hazards* 106(1): 435–457. https://doi.org/10.1007/s11069-020-04470-2.

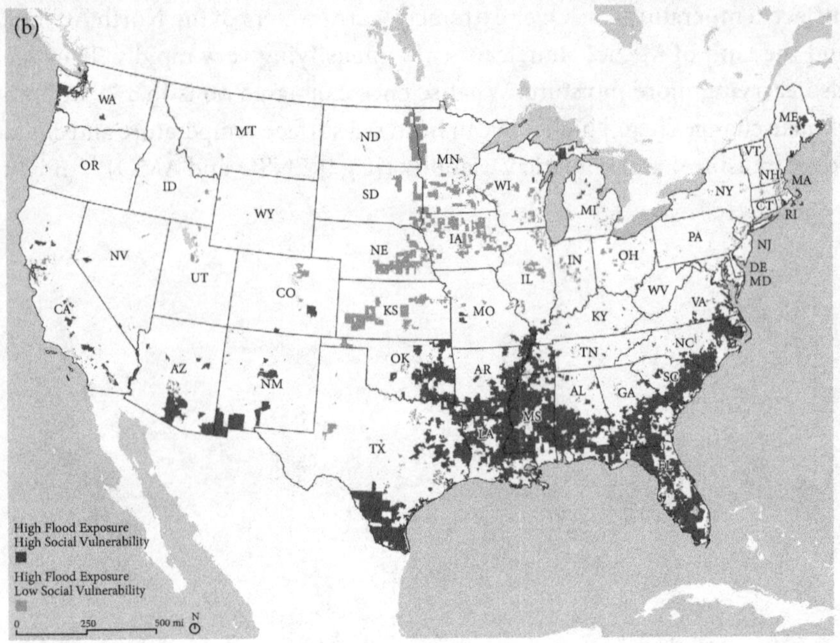

Map 2.3 Continued

explain the increase in the total number of tropical cyclones and major hurricanes in the North Atlantic."[185]

In terms of life safety, storm surge is the leading cause of death from hurricanes.[186] A combination of wind, wind-borne missiles and debris, waves, and storm surge pulls apart a building one exposed layer at a time.[187] In terms of economic losses, wind is the big problem from hurricanes and even tropical storms. In recent years, hurricanes and flooding together drove about $54 billion in losses every year, and wind damage alone accounted for $14 billion in losses.[188] As will be discussed in chapter 4, this is a major driver of insurance repricing and the emerging dysfunction of the wind insurance market.

The good news is that the building codes that require the securing of roofs, structures, and foundations are effective at reducing losses. For example, the Florida Building Code (FBC) developed after Hurricane Andrew has been very effective, with one estimate suggesting a 72% reduction in losses.[189] Despite consistent efforts by the state legislature to gut the FBC at the behest of the home building industry, a benefit-cost analysis of the FBC yields a benefit of six dollars for every one dollar spent in compliance with the code.[190]

Part of the FBC is organized around wind maps from the American Society of Civil Engineers (ASCE) that set the standards for coastal design and construction immediately on the coast. Not only must the structure withstand the winds, but it must also withstand the debris being blown into the building, as well as the larger pressure differentials (between the inside and outside of the building) that are created during wind events. The problem is that the maps have not kept up with increasingly more intense hurricanes that are hitting all parts of the state—from the Keys to the Panhandle. An ongoing update to these wind maps will mandate that certain classes of buildings withstand up to 89.4 mps (200 mph) winds.[191] As will be explored in chapter 4, Florida is one major hurricane away from a near-total failure of the insurance market, and this could very well be the breaking point for many future climigrants who simply will not be able to insure their homes, cars, boats, and valuables.

Extreme Heat, Droughts, and Wildfires

Climate change is super simple. The sun sends approximately 341 watts per square meter (W/m^2) of radiant energy to the earth. Some of this energy gets reflected back into space, and some of it is absorbed by the atmosphere, the oceans, and the land. Without some greenhouse gases in the atmosphere, the planet would be cold. Human-caused GHG emissions are responsible for retaining an extra amount of radiant energy—known as "radiant forcing"—that amounts to about 2.79 W/m^2.[192] A relatively small amount of extra energy trapped in the atmosphere has big effects on the climate and on temperatures. As GHG emissions go up, so too does planetary warming.

It is getting hot outside—extremely hot. From 2011 to 2020, global surface temperatures were 1.1°C (2.0°F) above observed temperatures from 1850 to 1900.[193] But, the United States is warming faster than the rest of the world—at least by the averages. As the *NCA5* noted, "Temperatures in the contiguous United States (CONUS) have risen by 2.5°F and temperatures in Alaska by 4.2°F since 1970, compared to a global temperature rise of around 1.7°F over the same period."[194] 2024 was the warmest year on record across the globe, and the year broke records for temperatures over the land and sea.[195]

Historically, measuring the temperature over the land and sea is complicated by measurement and instrument bias, with many more instruments in use over land and in well-resourced counties such as the United States. Today, however, computational models and satellites provide a great deal of precision. Although things like natural variability and volcanic activity play a role in global temperatures, human drivers are the overwhelming cause of increases in global temperatures.

Scientists still have a lot of questions. As the world begins to address climate change, sometimes unexpected things happen. For instance, in an effort to cut emissions from global shipping, regulations enforced new standards for sulfur content in fuels, which was effective in reducing emissions by approximately 80%.[196] What most people did not fully anticipate is that these aerosols were sending some measure of radiant energy (~0.13 W/m^2) back into space.[197] One immediate implication of these reduced aerosols was observed accelerated warming across the world's oceans, including in the Arctic and over the Atlantic.[198] This is potentially bad news for the East Coast of the United States to the extent that warmer sea surface temperatures in the Atlantic are more conducive to the development of tropical systems.

When it comes to global warming, most of the time people are just paying attention to the averages without fully appreciating the implications for daily life. The winters are warming twice as fast as summers in many states, which has implications for everything from precipitation to the management of drinking water infrastructure.[199] One concerning trend is that "nighttime temperatures are rising faster than daytime temperatures."[200] This is particularly problematic because buildings and urban landscapes increasingly do not have a chance to cool down overnight, which can amplify urban heat island effects.[201] This has potentially widespread health implications from the disruption of sleep[202] to premature mortality.[203]

Buildings were designed to balance energy coming in and energy going out. When that balance is thrown off, the thermal comfort of occupants can be negatively impacted. This impacts students trying to learn[204] and patients trying to heal.[205] Although air conditioning can help, extreme heat negatively impacts students' capacity to learn and their performance on standardized tests.[206] By one estimate, 36,000 public schools in the United States need to either procure or update their heating, ventilation and air conditioning (HVAC) systems.[207]

Although there are no firm national numbers, it is estimated that thousands of nursing homes do not have air conditioning, and, to make matters worse, there is very limited state legislation mandating the installation of air conditioning.[208] With an aging society and an increase in demand for nursing homes, many facilities in the Sun Belt without air conditioning will be dangerous. Unfortunately, many facilities are already dangerous in large and small ways. There have been dozens of fatalities at nursing homes from storm-driven power outages in recent years. An interview with a phlebotomist at a healthcare facility without air conditioning also noted that people are much more likely to faint on hot days after giving blood. For older people, the risk of losing consciousness raises the collateral risks of broken hips and other severe injuries. Retiring in the Sun Belt may be more dangerous than people realize.

Degree days are "the number of degrees by which the average daily temperature is higher than 65°F [18°C] (cooling degree days) or lower than 65°F [18°C] (heating degree days)."[209] As one might imagine, the number of cooling degree days has gone up, and the number of heating degree days has gone down since the 1980s. Unfortunately, it is estimated that about 1 in every 10 homes does not have air conditioning in the United States.[210] For many people who have air conditioning, they may not be able to afford to use it. It is not unheard of for people in the Sun Belt to pay more in the summer for electricity than they pay in rent.

Changes in heating and cooling degree days demonstrate a clear trend in the built environment, with broad implications for future projected energy use. An increase in extreme heat events is also a matter of life and death. Research highlights a direct correlation between increased temperatures and "demand for emergency services, including police and fire department incidents."[211] The increased utilization of air conditioning has its own set of unintended consequences associated with the indirect mortality of people from the air pollution associated from an increased demand for energy.[212] In full recognition that heat waves can also drive power outages,[213] there is increased interest in designing and regulating housing so that it can maintain the passive survivability of occupants when the air conditioning inevitably goes out.[214]

According to the U.S. Global Change Research Program (USGCRP), the "average number of heatwaves has doubled [in the United States] since the 1980s, and the length of the heatwave season has increased from about 40 days to about 70 days."[215] Map 2.4 highlights the projected increase in the number of days over 38°C (100°F). As heat waves become more frequent, people pay the price. Some people are merely inconvenienced by

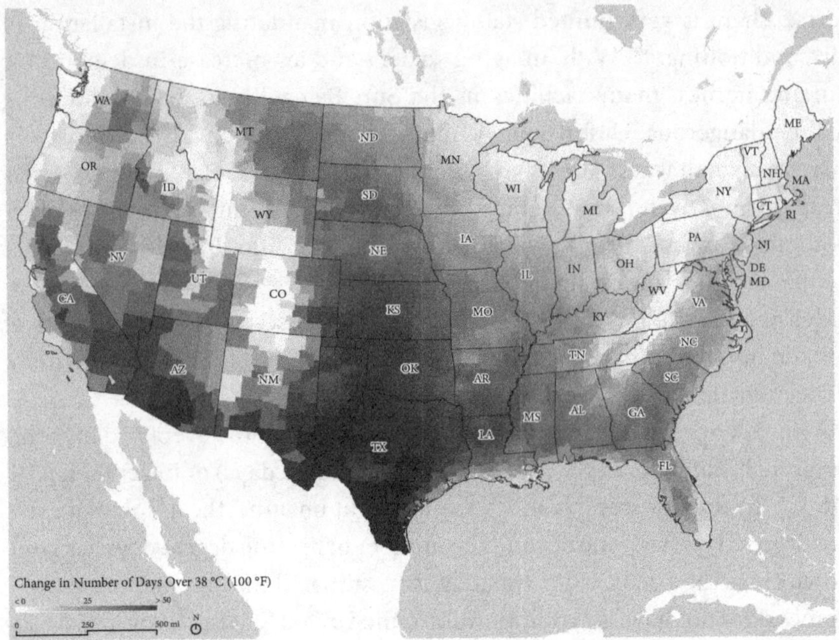

Map 2.4 Projected Change in Number of Days over 38°C (100°F) from 1991–2021 Baseline with 3°C (5.4°F) of Global Warming. Design Credit: Oliver Oglesby and Jesse M. Keenan Data Source: U.S. Global Change Research Program. 2023. National Climate Assessment: Interactive Atlas Explorer. Washington, D.C.: U.S. Global Change Research Program.

transportation and mass transit delays, from everything from buckled rail lines to power outages.[216] Others are not so lucky. More than 1,200 people are killed by extreme heat events every year in the United States, as either an underlying or contributing cause.[217]

A warming world is also driving drought. The *NCA5* noted that "drought results when there is a mismatch between moisture supply and demand. Meteorological drought happens when there is a severe or ongoing lack of precipitation. Hydrological drought results from deficits in surface runoff and subsurface moisture supply."[218] Although the levels of uncertainty around drought projections vary by region in the United States, the general arch of the research is that the risks are increasing.[219] The implications of drought on the built environment are vast. The most obvious challenge relates to water insecurity. Water is key not just for human survival, but it is also key for supporting economic output[220] and energy production.[221]

Energy demand also increases because water has to be pumped from deeper depths and over farther distances. In addition, droughts can destabilize soil, which impacts building foundations.[222] Airborne dust can reduce indoor air quality (IAQ) and reduce the effectiveness of HVAC systems in buildings.

People need drinking water, and without it, the adaptation options are limited beyond extreme conservation, desalinization, and long-distance pipelines. Map 2.5 highlights a projection for the future climatic water deficit facing the United States by mid-century. In many places, a lack of water is already shaping how and where people live. For instance, the moratorium on new housing development in Phoenix in 2023 is likely just the beginning of the challenge that developers will face in demonstrating that their projects will have sufficient long-term access to water.[223] By contrast, some jurisdictions, such as Kyle, Texas, lack the legal capacity to manage

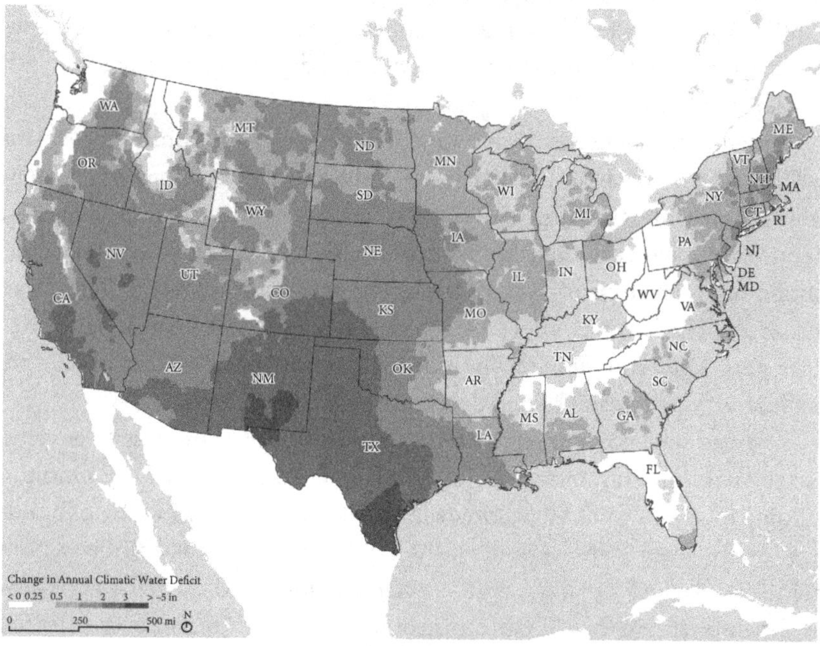

Map 2.5 Projected Change in Annual Climatic Water Deficit 2036–2065 from 1991–2020 Baseline. Design Credit: Oliver Oglesby and Jesse M. Keenan. Data Source: Payton, E. A., A. O. Pinson, T. Asefa, L. E. Condon, L.-A. L. Dupigny-Giroux, B. L. Harding, J. Kiang, D. H. Lee, S. A. McAfee, J. M. Pflug, I. Rangwala, H. J. Tanana, and D. B. Wright. 2023. "Ch. 4. Water." In Crimmins et al., *Fifth National Climate Assessment.*

growth.[224] As the second-fastest-growing U.S. city in 2023, it has so little water that the soil underneath peoples' homes is drying out and causing the foundations to sink and crack.[225] It is not uncommon for entire subdivisions and communities in the West and Southwest to have water trucked in weekly, at a great expense.[226] Not only does this trucked water increase fossil fuel consumption; the weight of the trucks accelerates the deterioration of roads.[227] Interviews highlight that those who cannot afford shipments of potable water run the risk of a public health condemnation and a default on their mortgages.

The West, Mountain West, and Southwest have a much bigger near-term problem—climate-fueled wildfires. When it comes to wildfire smoke, it is everyone's problem. In certain regions, increased atmospheric aridity draws increased levels of moisture from the biomass and soil, and this creates more fuel for the fire.[228] If one adds millions of dead trees killed from the pests and diseases with expanding ranges,[229] the result is a combustible mix that is difficult to manage.[230] Map 2.6 highlights the projected change in days with extreme weather conditions (e.g., hot and dry) that are associated with very large wildfires.

Although the science is undeveloped, in theory more convective storms could create more lightning that may ignite more fires. The *NCA5* highlighted that "while [wild]fire risk is not solely determined by climate factors, the authors have *very high confidence* that the hot and dry weather conditions that elevate fire risk are becoming more common."[231] The more immediate challenge is people accidentally or intentionally starting fires during an expanding fire season. Indeed, the human dimension of wildfire risk is largely a function of bad decisions.

Local and state governments have allowed people to continue to develop into the wild-urban interface (WUI). One nationwide study of 125 million building locations found that somewhere between 5.6% and 18.8% of buildings in any given state were in the WUI.[232] Although not entirely within their control, the federal government has also looked the other way; housing development in the WUI in and around National Forests has grown 46% from 1990 to 2010.[233] Overall, the growth in the WUI has been remarkable, with "50 million homes . . . currently in the [WUI] in the United States, a number [that is] increasing by 1 million houses every 3 years."[234] Across the country from 2000 to 2018, 54,000 buildings were burned, and upward of 97% of these buildings were in the WUI.[235] It is not just forests that are

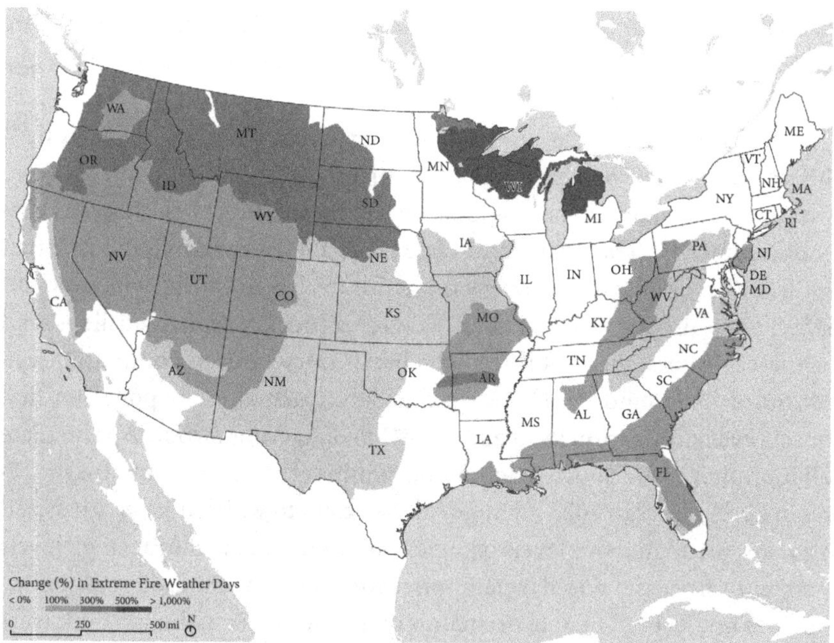

Map 2.6 Projected Change (%) in Extreme Weather Conditions Associated with Large Wildfires During Fire Season (May to October) for 2040–2069 from 1971–2000 Baseline Under RCP 8.5. Design Credit: Oliver Oglesby and Jesse M. Keenan. Data Source: Domke, G. M., C. J. Fettig, A. S. Marsh, M. Baumflek, W. A. Gould, J. E. Halofsky, L. A. Joyce, S. D. LeDuc, D. H. Levinson, J. S. Littell, C. F. Miniat, M. H. Mockrin, D. L. Peterson, J. Prestemon, B. M. Sleeter, and C. Swanston. 2023. "Ch. 7. Forests." In Crimmins et al., *Fifth National Climate Assessment.*

burning. There is also an increased risk of fire from grasslands and shrublands, as evidenced by recent devastating fires in Maui and Colorado.[236]

When wildfires strike, they can burn communities to the ground and kill people in the process. Thousands of people die every year—mostly from smoke exposure—from wildfires,[237] and more than 400 wildland firefighters have been killed in the line of duty since 2000.[238] Roads and bridges buckle and melt. Devastation is often total. Although building code modifications that impose things like fire-resistant building materials (e.g., roofing shingles), designs (e.g., metal screens to keep out flammable bird nests from under the roof), and landscape buffers have been shown to provide remarkable private and public benefits,[239] these marginal reductions in risk are often up against increasingly large and fast-moving fires that are more destructive, as evidenced by "a 160% higher structure-loss rate (loss/kha burned) over the past decade [2010–2020]."[240]

There is also the lasting impact of toxic contamination in the soil.[241] This comes from everything from the burning of building contents (e.g., furniture, household cleaners, appliances) to the melting of plastic water distribution lines.[242] This toxic landscape represents a fundamental risk that significantly complicates recovery in the future. To add insult to injury, climate change also increases the risk of extreme precipitation, landslides, and flash flood events after wildfires, which can bury burned communities in a slurry of toxic mud.[243] The toxic exposure extends to air pollution that impacts everything from localized IAQ[244] to dangerous airborne particulate pollution that is transported across the continent.[245] The devastating wildfires in Los Angeles in 2025 highlighted the extent to which fires in urbanized areas can spread massive levels of toxic contamination, including high levels of titanium from the popular paint pigment titanium white. This air pollution does not just kill people prematurely; it gives them asthma,[246] it lowers birth weights,[247] and it even drives some people to suicide.[248]

There is no escaping the impacts of wildfires, with New Yorkers and Duluthians often bearing the health consequences of wildfires in California and Canada. Many people find these impacts are mentally overwhelming. Indeed, some people may be inclined to decompress with a glass of wine after reading this chapter. Just make sure that it is not a California vintage whose smokey flavor is an outcome of the countless toxic contaminants that the grapes have absorbed from the burned remains of peoples' homes.[249]

There is no escaping climate change. Both sending and receiving zones across the United States will face a range of climate impacts. Intervening extreme events in receiving zones and climate destinations may entirely change the amenity value of these places for prospective climigrants. Although no place is immune, some places will be better off than others. For the Sun Belt, declining environmental amenities and increased risks are converging to alter the migration calculus for generations to come. As this migration pattern reverses, the Northeast and Midwest stand to inherit a new generation of climigrants eager for "nice weather," affordable housing, and a built environment that is adapted to the impacts of a changing climate.

3

Climate Adaptation Science, Policy, and Planning

Throughout the course of multiple periods and cycles of climate change, humanity has always adapted. When societies and civilizations do not sufficiently adapt, they fail. When biological life and non-living viruses do not adapt, they go extinct. Adaptation is a complex range of processes that can be contextualized to different stimuli, over different periods of time, by a wide range of actors and agents. This chapter seeks to bridge a biological and ecological understanding of adaptation processes with the human adaptations that shape the preparations for and responses to climate impacts. Along the way, allied concepts such as resilience, maladaptation, and transformation are explored under a broader umbrella of adaptation science, policy, and planning. The intent is to narrow this broad field of multidisciplinary science to a range of considerations that resonate within the arch of the built environment, public policy, and population mobility.

Climigration and relocation are themselves a form of adaptation. In the context of how and where people will live, adaptation policies and plans (APPs) are rapidly organizing interventions across sectors and geographies in the United States. The resulting public and private investments are likely to shape a range of conditions that will speak to who gets protected and who is left behind. This chapter explores these emerging APP activities in order to provide insight into the capacity of Americans to adapt their homes, communities, infrastructure, and institutions to a range of climate impacts outlined in chapter 2. When people get left behind, markets and infrastructure fail, and institutions unwind, then the specter of maladaptation is always on the horizon. As one senior federal official once said, "The goal is not to define or drive adaptation. The goal is to prevent maladaptation."

North. Jesse M. Keenan, Oxford University Press. © Oxford University Press (2025).
DOI: 10.1093/9780197641644.003.0004

Climate Adaptation Science

Adaptation science is somewhere between an interdisciplinary and transdisciplinary endeavor to understand the nature and mechanics of change. To understand how people adapt, one must first appreciate how the human species fits into a broader realm of biological and ecological adaptation science. The negotiations among the physical, biological, and social sciences are often fraught with miscommunications and metaphors. Yet common features associated with synchronous and asynchronous energy and resource flows dictate a shared path in the face of climate change, a biodiversity crisis, and continuing environmental degradation. Outlining this shared path is the basis for understanding interventions advanced in the name of both sustainability and adaptation.

Adaptation in Biology and Evolution

At the very heart of evolution is the idea of adaptation. Through *phenotypic adaptation* species adjust their behaviors and physical features to advance their fitness in a dynamic and changing environment. This is sometimes also referred to in the context of acclimatization, where species are trying their best to survive in a new or modified climate. Over time, *genotypic adaptation* arises from genetically encoded changes that are both functional and increase the fitness of the species to its environment. As will be discussed, where there is adaptation, there is also the risk of maladaptation. In this context, maladaptation might result in the development of vestigial features, which are largely non-functional or become non-functional over time. While phenotypic responses to climate change are reasonably observable for many species (e.g., range shifting), genotypic changes are more difficult to measure given the relatively short time that anthropogenic climate change impacts have been felt.[1]

While fruit flies (*Drosophila melanogaster*) get all of the fanfare for being the first species with evidence of climate-driven genotypic adaptation,[2] the pitcher-plant mosquito (*Wyeomyia smithii*) has the distinction of being the first animal to have its genetically encoded evolutionary changes to climate change actually mapped with precision.[3] This genotypic adaptation is a consequence of different post-glacial pitcher-plant mosquito populations at different latitudes and altitudes sharing genetic material with populations that have successfully adapted in warmer, sunnier, and wetter geographies—in this case, the Gulf Coast and the Carolinas.[4] The ability of mosquitos to so quickly adapt to climate change is not good news in terms of the spread of vector-borne diseases.

One debate in the science revolves around *phenotypic plasticity*, which is "the expression of multiple phenotypes from one genome," and the extent to which it limits or "facilitates evolutionary adaptation to climate change."[5] Indeed, the answer might be different for different species. Emerging research suggests that "genetic variation in plastic responses to the environment . . . might be an important predictor of species' vulnerabilities to climate-driven decline or extinction."[6] This underscores the value, for species like fruit flies and mosquitos, of possessing a great deal of genetic and population diversity. Diversity is a key element for many processes of adaptation. Without such diversity, extinction is a real threat for species. The Bramble Cay melomys (*Melomys rubicola*)—a kind of rat—has the unfortunate distinction of being the first mammal to go extinct from climate change.[7] The melomys lived on low-lying islands off the coast of Australia. Storm surge and SLR rise were just too much for the rats. By one estimate, somewhere between 16% and 30% of 538 major plant and animal species could go extinct from climate change by 2070.[8]

For decades, research has shown that some species adapt by changing where they live and reproduce along different seasonal, altitude, and latitude gradients.[9] Map 3.1 highlights the extent to which climate change is projected to significantly shape the ranges and habitats of species, with many species in the Northern Hemisphere projected to be moving north. This is nothing new. There is robust palaentological evidence that species have changed latitude in response to previous changes in climate.[10] In the Northern Hemisphere, this often means moving, flying, and swimming north for cooler temperatures, more nutrients, or other attractors that are key to a species' ecosystem fitness. Simply moving from one habitat to another is easier said than done. Species live within *complex adaptive systems*, as part of a web of interacting systems and SoSs whose macroscopic behavior is in a constant state of dynamic interaction and reorganization.[11] All species rely on other species to do things that support their survival. When species seek to adapt by moving somewhere new, this synchronicity of time, place, and function can break down. The resulting asynchrony puts stress on species and ecosystems, and it may even shorten lifespans.[12]

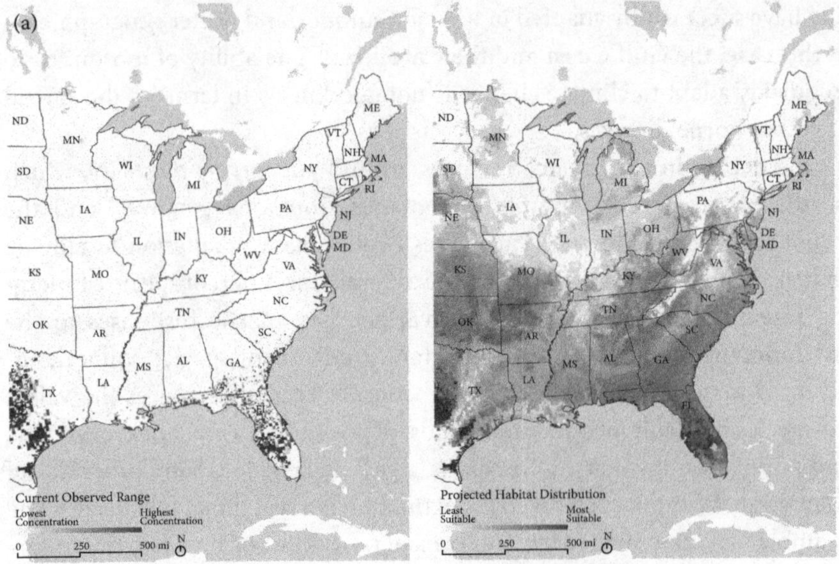

Map 3.1 Comparison of Current and Projected Range Characteristics Under RCP 8.5: (a) Live Oak (*Quercus virginiana*); (b) Blue Jay (*Cyanocitta cristata*). Design Credit: Oliver Oglesby and Jesse M. Keenan. *Data Sources*: (a) Northern Research Station, U.S. Forest Service. 2024. "Climate change atlas: Tree atlas (version 4): Live Oak (*Quercus virginiana*)." Madison, WI: U.S. Department of Agriculture; (b) Northern Research Station, U.S. Forest Service. 2024. "Climate change atlas: Bird atlas (version 2): Blue Jay (*Cyanocitta cristata*)." Madison, WI: U.S. Department of Agriculture

An Ecological Perspective on Adaptation and Resilience

With or without climate change, natural variation always drives dynamic periods of stress on species and ecosystems. In theory, complex adaptive systems reorganize in the face of such changes. In ecological terms, the *adaptive cycle* has been a useful concept for describing the movement through space of a species and/or elements of an ecological system. The adaptive cycle is defined by the stages of exploitation, conservation, release, and reorganization of capital.[13]

From here, the concept of *panarchy* describes the constant state of movement of capital by and between different nested adaptive cycles.[14] It is the flow of capital across the cycle of life and across nested SoSs. What scientists

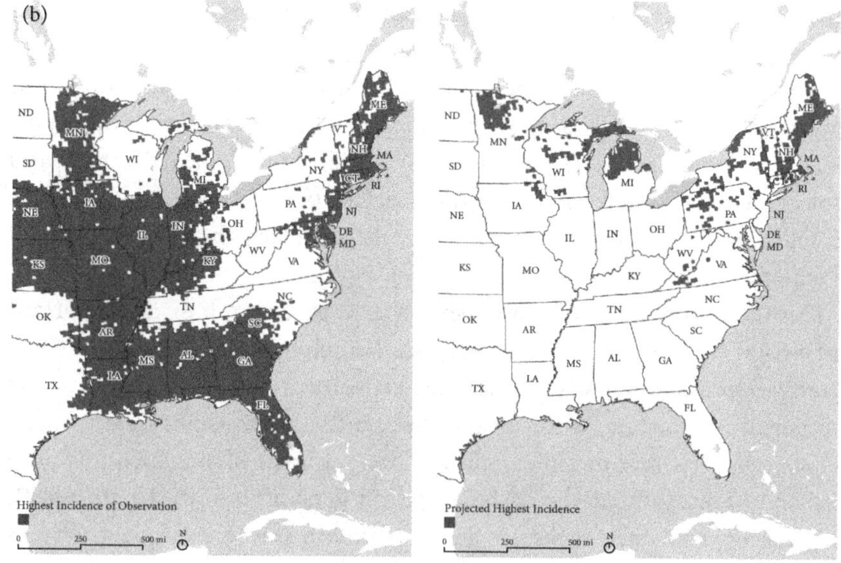

Map 3.1 Continued

observe at any moment is a mere reductionist snapshot of this flow of capital, as nothing is inherently stable.

The adaptive cycle offers ways of describing this flow of capital across ecological systems. System potential is "concerned with the range of options available for future responses of the system."[15] Connectedness speaks to the extent that system elements "are dominated by external variability" in a manner that shapes the relationships and interactions among system elements and processes.[16] Finally, and most importantly, *ecological resilience* is the "ability to absorb change and disturbance and still maintain the same relationships between populations or state variables."[17] For example, ecological resilience might be a measure of how much of a shortage—of particular food that is central to the diet of a particular species—can be absorbed before a species is forced to find new sources of nutrition in new places, at new times, or even at different times of year. Once the threshold of resilience is crossed, species either adapt or starve—or even go extinct.

For the pioneering ecologist C. S. Holling, the single evolutionary goal of species was not stability but the avoidance of extinction. He argued that early ecological models were deterministically shaped by the anthropocentric goal of maintaining populations and managing resources.[18] But nothing in nature is stable. Holling's idea of resilience was centered on a measure of the "persistence of relations," which may absorb perturbations and allow ecosystems to continue with their core ecological processes or functions. For Holling, the persistence of these relationships—not necessarily the persistence of a particular species—is driven by flexibility, biological diversity, and ecological heterogeneity in time and space. He observed that highly unstable ecologies, such as forests that regularly burn, often have the greatest capacity to maintain resilience.

This influential body of scholarship would go on to draw an important distinction between ecological resilience and engineering resilience. *Engineering resilience* is concerned with the capacity of a system to return to its pre-perturbation state, with the assumption that there is only a single stable state.[19] This is sometimes referred to as a form of single-equilibrium resilience. For example, the design of power generation infrastructure with engineering resilience measures helps maintain the flow of electricity in the face of disruptions. There is only one single stable state—when the power is on.

By contrast, ecological resilience assumes that there is not just one single stable state, but rather there are multiple potential stables states.[20] Perturbations (disturbances, shocks, extreme events) can shift ecosystems from one stable domain to another, but the general identity of the system is largely intact under a theory of ecological resilience. This is referenced as a form of multi-equilibrium resilience.[21] Ecological resilience is centered on the processes that keep a particular ecological system within one of these multiple potential domains. In practice, it is indirectly measured by a variety of ecosystem health and environmental quality indicators.

In anthropocentric terms, a system identity may operate as a metaphor for a range of factors that shape stable states. As such, the stability of a human construct is not always a good thing, "[as it] may prove very difficult to transform a resilient system from the current state into a more desirable one."[22] As will be discussed later in this chapter, there is no single form or type of resilience, and not all resilient states are desirable. Rather there are multiple types of resilience that each have their own conceptual, analytical, and empirical foundations.

All types of resilience have a threshold. When that threshold is crossed, one either adapts or one fails. But what does it mean to adapt? The answer to this question requires a somewhat abstract frame of reference, as it is not always clear whether something has adapted or simply adjusted on the margins. In general, everything, from ecological systems to people, has a contextual *identity*. That identity is grounded in rules and relationships over time and space.[23] For people, that might be relationships associated with dependent physical infrastructure or social relationships defined by social and cultural norms.[24] Although not everything is part of a system[25] and not all systems are functional,[26] this somewhat abstract concept of identity is what defines a system or person's functional existence.[27] Both engineering resilience and ecological resilience are working to maintain that identity, even though some of these rules and relationships might change in response to an external perturbation.[28]

When the thresholds of resilience are crossed, then the object's identity either collapses or it transforms into an entirely nonlinear and novel set of rules and relationships that reflect a fundamentally new identity with its own contextual state of existence and performance. Although there is some scholarly debate around the uneasy conceptual merger of ecological and social processes,[29] this process of a fundamental change in structure and attributes is generally regarded as *transformative adaptation*.[30] Absent such fundamental changes, adaptations that largely maintain the existing identity of a system are considered *incremental adaptation*.[31]

The Adaptations of People

People have organized themselves into very complex societies, cultures, and governance bodies through everything from language and social norms to institutions and laws. Sometimes people engage in *planned adaptation* in anticipation of future events. In other cases, *autonomous adaptation* arises from a deficit of self-organizing behavior, belief systems, or signal intelligence to prepare, and, therefore, the best that people can do is to adapt after the fact. These regimes of social existence are in dynamic interaction with—albeit not necessarily in dynamic communication with[32]—meta-environmental and complex adaptive systems across a panarchic spectrum of chaotic relationships. Proclaiming that any unit of analysis within these complex socio-ecological systems has adapted may be a tall order to empirically defend. In the context of social behavior and complex institutions, part of the challenge is that transformative adaptations—and all types of adaptation—are processes and not outcomes.[33] In this sense, it is sometimes difficult to distinguish a post-transformative state, where adaptation processes have normalized, from an interim state of adaptation where the sought-after stable state is still unrealized or even fully foreseeable.

This is complicated by the fact that there are different types of change, including linear developmental change, which takes place incrementally over time; transitional change, where the focus is on the interim state; and true transformational change that represents a discontinuous alteration of identity and disconnection from the present state. It is hard to know where one is at any given moment. What may look like adaptation at one scale may really be maladaptation at another scale.

The IPCC defines *maladaptation* as "actions that may lead to increased risk of adverse climate-related outcomes, including via increased greenhouse gas (GHG) emissions, increased or shifted vulnerability to climate change, more inequitable outcomes, or diminished welfare, now or in the future."[34] Almost any adaptation presents the risk of maladaptation.[35] Trees planted for shade may damage buildings in a storm. Desalination plants that produce expensive water may disproportionately reduce access to low-income customers, who may drink less water and become dehydrated. Ski resorts adapt to changing alpine winter temperatures by using artificial snow, which increases GHG emissions. Successful adaptation by some may be an absolute maladaptive failure for others.

Conversely, what looks like failure from one perspective may actually be the beginning of a broader transformation. In a classic debate published in the journal *Nature Climate Change*, Dow et al. argue that adaptation is not unlimited because there are finite resources to support adaptation.[36] In this sense, adaptation has a frontier much in the same way that ecological resilience has a threshold. The authors highlight the temperature sensitivity of rice yields to climate change and the extent to which future temperature increases represent a clear limit to the adaptation of rice farmers in Southeast Asia. No matter what farmers do, rice yields will be too low to sustain farming. Rickards responds to Dow et al. by arguing that the failure of rice farmers might be the beginning of a broader positive transformation wherein the farmers might have otherwise been inefficient, and they may actually increase household wealth with higher-productivity non-farm labor.[37] Indeed, the reforestation or afforestation of land formerly used for rice cultivation might actually promote carbon sequestration.

This debate about the limits of adaptation led to the recognition that there are hard and soft limits, where "[adaptation] options may exist but are not currently available."[38] Soft limits are seemingly universal given the economic costs of adaptation and the technological limits of avoiding all the impacts of climate change. Adaptation is rarely free. Likewise, the distributional costs and benefits of adaptation are rarely in line with what policy makers might consider fair or just. Unfortunately, adaptation is often advanced by people and institutions in power who have little regard for collective goals.[39] In this context, one can argue that adaptation can really be understood only with the hindsight of history and the benefit of a removed understanding of the totality of complex economic and social processes driving adaptation. What the farmers in Southeast Asia might describe as a devastating failure of their livelihoods, historians might describe as the transformative adaptation of agricultural ecosystems and economies.

This historical perspective is of little use in light of the immediate demands to plan for and invest in adaptation. In terms of contemporary APP, people are challenged to address the cross-scalar subjectivities of adaptation within the context of the desire to objectively measure successful adaptation itself.[40] The subjectivities are centered on two fundamental questions: (i) Adapt to what? and, (ii) Who is adapting?[41] Often adaptation processes are subjectively elusive. Sometimes the best that planners can do is to either try to understand the path dependencies that are created by current decisions and behaviors[42] or try to measure and promote the capacity of people to adapt. This latter concept of *adaptive capacity* is defined as the "ability of systems, institutions, humans and other organisms to adjust to potential damage, to take advantage of opportunities or to respond to consequences."[43] The concept can even be extended to buildings and infrastructure systems that can be designed to accommodate future unknown environmental conditions and programs.[44]

In defining *adaptation*, the IPCC draws a crucial distinction: "In human systems, [adaptation is] the process of adjustment to actual or expected climate and its effects, in order to moderate harm or exploit beneficial opportunities. In natural systems, [it is] the process of adjustment to actual climate and its effects; [wherein] human intervention may facilitate adjustment to expected climate and its effects."[45] This widely recognized definition is key for two reasons. First, it emphasizes that people are managing risk. Aside from conventional material or structural interventions, these adaptations may include engaging natural systems through ecosystem management activities to facilitate ecological adaptation, increase engineering resilience, and reduce overall social vulnerability. This is often referenced as *ecosystem-based adaptation* (EBA).

Second, and this is key, adaptation may also be advanced to take advantage of opportunities. Climate change and the elements of global change (e.g., aging societies, artificial intelligence) are going to rewrite the rules that define society and the global world order. Rewriting these rules in a manner that advances broader long-term societal goals, such as sustainability, is an important element of adaptation. Specific to this book, climigrants and people who are relocating have an enormous opportunity to advance a sustainable footprint in the built environment of wherever they end up. More fundamentally, they have the opportunity to redefine an American way of life that provides the basis for a more environmentally responsible, humane, and procedurally fair country.

In the near term, people will be tasked with the immediate challenges of coping with and adapting to climate change impacts. Indirectly, this behavior also includes adaptations that are responsive to the *transition risks* that come along with decarbonizing economies and societies. For instance, fossil fuel-dependent communities will have fewer resources in the future to adapt, and this represents a unique challenge it in its own right.[46] While transition risks are important, the primary goal of adaptation is associated with the moderation of the physical risks outlined in chapter 2.

In this light, successful adaptation might accomplish one of three things.[47] First, it could reduce the *sensitivity* of a system, object, or species by reducing the "degree to which a system or species is affected, either adversely or beneficially."[48] For instance, a building with water resistant materials is much less sensitive to a flood than a building with conventional materials. Second, adaptation can alter the *exposure* of a system, insofar as people, species, or objects are physically located in places where they may be impacted. When a community relocates out of an SLR zone to an area of higher elevation, they have reduced their direct—but maybe not their indirect—exposure to SLR. Finally, successful adaptation may increase people's or even a system's adaptive capacity in the future.

Figure 3.1 provides a schematic of the major distinctions between resilience and adaptation, as well as the closely related concept of risk mitigation. In short, *risk mitigation* is about preventing an external perturbation from impacting an object or system. For example, a flood wall is a form of risk mitigation because it is attempting to prevent the water from even reaching a building or community. In its most simplistic form, adaptation is about transforming to an entirely new domain of identity and performance, often through changes in exposure and sensitivity. By contrast, resilience, notably

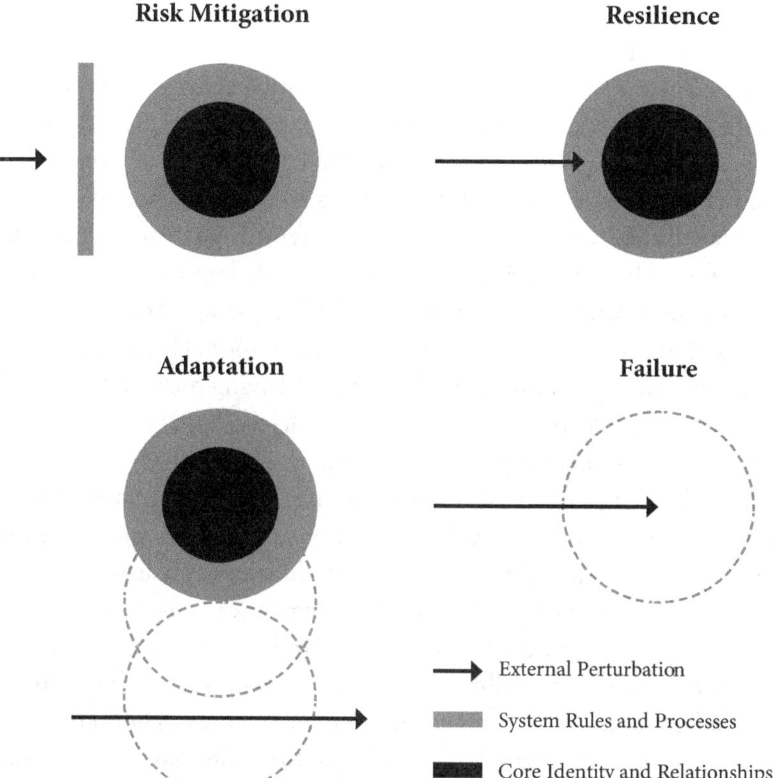

Figure 3.1 Conceptual Distinctions Among Risk Mitigation, Resilience, Adaptation, and Failure. Design Credit: Oliver Oglesby and Jesse M. Keenan

engineering resilience, is fundamentally about the capacity of an object or an ecosystem to absorb a perturbation and maintain its core identity and performance.

Making Sense of Resilience

Resilience is one of the most grossly misunderstood concepts in climate discourse. As has been previously discussed, not all human interventions advanced in the name of resilience are positive. This is also true for adaptation, where the risk of maladaptation is always present, often as an unintended consequence. In practice, the risk of maladaptation with resilience is often elevated. This is because when most people talk about resilience, they are talking about engineering resilience.[49]

Engineering resilience is arguably a conservative concept because it conceptualizes only one stable state. Formally, engineering resilience processes drive an elasticity to revert to the status quo. The problem is that the status quo is not always a great place to be. As some scholars have observed, "There can also be a negative dimension of having high resilience. A system can sometimes become resilient in a less desirable regime. . . . [R]esilience can work as both a vehicle of sustainability and [as] an agent of destitution."[50] By implication, resilience can reinforce the institutional lock-in of rules, as well as established power regimes that may be driving maladaptation.[51] For instance, flood protection systems are often driven by large and powerful engineering conglomerates who would much rather build steel and concrete structures than engage in low-carbon EBA or processes of managed relocation that may offer superior social welfare outcomes. This phenomenon is sometimes referenced within the context of "disaster capitalism."[52]

Engineering resilience can be a critical element of conservation, preservation, and physical protection. As a single-equilibrium process, it can also work against the multi-equilibria that may be achieved through different types of adaptation.[53] For instance, massive investments in flood protection systems that serve an engineering resilience function may actually undermine the economic capacity for communities to engage in more effective non-structural approaches, such as through managed relocation or community relocation. Engineering resilience can also become overspecialized with a focus on one risk or hazard, while overlooking contextual risks that might be just a damaging.[54] For example, after the 2011 Tōhoku earthquake in Japan, the government sought to build resilience to tsunamis by moving communities to higher elevations on the hillsides.[55] The downside was that the risk of landslides from flash floods on these hillsides was much greater than the risk of another tsunami. In other cases, resilience and adaptation may be competing for the same resources. In coastal Australia, it was observed that wealthy property owners immediately on the coast often wanted to spend public dollars on resilience to protect their properties, while other stakeholders were more attuned to the necessity to shift investments away from coast, as form of adaptation.[56] Resilience can be a double-edged sword.

Although the contemporary popularization of the concept of resilience can be traced to the ecologist C. S. Holling, the word itself can be traced to the

Latin term *resilere* (*resilio*), which means "to bounce."[57] The term appears in the works of Seneca the Elder, Ovid, Cicero, and Livy, who used it in varying ways to describe leaping or rebounding. The root of the word can be traced through the Romance languages and would even appear in the State Papers of King Henry VIII in his attempt to extricate himself from his troubled marriage with Catherine of Aragon.[58] In the nineteenth century, the term resilience, or its derivative resiliency, would be used to characterize Edo's (Tokyo's) recovery from an earthquake in 1854 and would be widely used in mechanics, surgery, anatomy, and watch making to describe the elastic properties of materials.[59] In the twentieth century, resilience concepts would be extended to understand the psychological coping mechanisms of children and adults to everything from alcohol abuse to organizational changes in the workplace.[60]

Today, multiple types of resilience are drawn from a variety of disciplines and practices. Figure 3.2 highlights a non-exclusive range of resilience types that are widely utilized in policy and planning. For the most part, engineering resilience, ecological resilience, and community resilience represent the most well-developed types of resilience utilized in practice in the United States. These types of resilience can be understood to exist along several conceptual continua. First, as previously referenced, some types of resilience operate to revert to a single-equilibrium state. This is most common in engineering resilience and *disaster resilience*, which is a related form of engineering resilience that is more narrowly applied to the dependent interaction between people and infrastructure systems within the context of natural hazards and disasters. Because there is just one stable state, the processes of resilience are descriptive insofar as the post-perturbation reversionary processes can be measured in terms of time, performance, and cost.

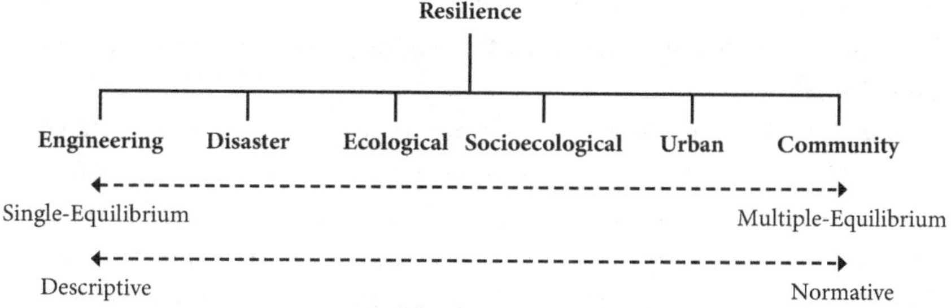

Figure 3.2 Major Types and Characteristics of Resilience. Design Credit: Oliver Oglesby and Jesse M. Keenan

Other forms of resilience, such as ecological resilience, exist in the context of multiple potential future states. These types of resilience tend to be normative to the extent that there is a tendency to design indicators that measure what people think would or should define a desirable potential future state.[61] While *ecological resilience* is often measured according to the perturbation and not the elastic response, there are normative indicators for ecological resilience that speak to fundamental attributes of environmental quality.[62] In biological and ecological terms, the idea of multiple future states makes sense because of the self-organizing behavior of chaotic complex adaptive systems. As such, the evolutionary processes of adaptation may result in any number of adaptive outcomes, particularly in the context of species acclimatization that may be limited only by phenotypic plasticity.

Community resilience—sometimes referred to as social resilience—is another form of multiple-equilibrium resilience found in policy and planning. Community resilience is defined as

> the ability of communities to withstand, recover, and learn from past disasters, and to learn from past disasters to strengthen future response and recovery efforts. This can include but is not limited to physical and psychological health of the population, social and economic equity and well-being of the community, effective risk communication, integration of organizations in planning, response, and recovery, and social mobilization for resource exchange, cohesion, response, and recovery.[63]

Multiple potential future states are defined by communications, discourse, and the capacity for social learning. As people experience not only disasters but also the chronic stresses of climate impacts, they learn and share different aspects of that experience.[64] Over time, this shared experience and accumulated knowledge help build social, financial, and other forms of capital that drive collective goals for risk mitigation, recovery, and even adaptation.[65]

There is a limit to how far community resilience can be extended. From 2015 to 2025, many communities experienced post-disaster planning fatigue, particularly around the concept of resilience. Many communities in high-risk areas were subject to the widespread indoctrination of an undisciplined framing of resilience by groups such as the Rockefeller Foundation.[66] When and if things failed, communities often felt like it was their fault that they had failed due to their inherent shortcoming in building community resilience. As has been noted, "Critics of neoliberalism have [argued] that resilience—with its techno-managerial focus—has clouded the necessity for engaging collective responsibility and instead has focused on individual vulnerability."[67] By extension, resilience tends to focus on the individual or the community, and it often fails to drive transformations on higher-order institutions that define people's vulnerability in the first place. As scholars have noted, "If collective alternatives are not sought, the existing institutions that have contributed to such predicaments not only remain unchallenged, but [they] are relied upon to steer societal responses based on the same underlying assumptions that first led to the problems."[68] There are not only thresholds to resilience, but there are also limits to its value as a concept.

Understanding the Trade-Offs Between Adaptation and Resilience

Whether it is community resilience, engineering resilience, or any number of types of adaptation, trade-offs exist across actor orientations, time horizons, and geographies. The costs and the benefits of such interventions may have huge variations in terms of net distributional burdens. Adaptation by one person may increase vulnerability for another person. For instance, building flood protection systems on ocean-front properties might be driving engineering resilience for the property owners. Yet it might be maladaptive for their neighbors to the extent that flood walls might steer floodwaters and storm surge to neighboring properties. Over the long term, the interventions might even be maladaptive to the ocean-front property owners themselves, as the steered floodwater might impact infrastructure that they rely on. Therefore, it is incumbent on policy makers and planners to consider relevant actor orientations, time horizons, spatial boundaries, and trade-offs that might exist between different goals associated with different types of resilience and adaptation.

Table 3.1 provides an outline of a range of potential perspectives and trade-offs associated with the managed relocation of communities in the context of climigration. For instance, climigration may be a form of adaptation at the household level. But it may have maladaptive implications for immobile populations who are left residing in sending zones. This resident population may have fewer public resources for engineering resilience interventions and fewer measures of social capital to support community

Table 3.1 Trade-Offs Associated with Interventions That Support Managed Relocation and Climigration

Actor Orientation	Time Horizon	Adaptation (Actor Specific)	Maladaptation (Actor Specific)	Engineering Resilience	Community Resilience	Ecological Resilience
Local Households	Short Term or Long Term	Climigrants reduce risk exposure and may have greater opportunity in receiving zones.	Some people will be immobile and will be stuck in high-risk sending zones.	Climigrants may have a lower tax burden in receiving zones that do not have the same infrastructure cost structure as sending zones.	Community resources for mutual support may be undermined with out-migration.	Declining populations may reduce environmental stress.
Sending Zone Community	Current Generation	Climigrants have the opportunity to develop new forms of social capital in receiving zones.	Community is fractured and segmented by climigration and differences in wealth.	Greater reliance on engineering resilience interventions that may be limited in their capacity to adapt to new climate risks.	Community resources for mutual support may be undermined with out-migration.	Uncertain ecological impacts with fewer people, but also fewer resources for local environmental management.
Sending Zone Local Government	Annual Tax Assessment & Bond Terms	Infrastructure obligations decline in sending zones	Diminished tax base reduces opportunities for investing in future adaptation.	Declining fiscal resources limit the capacity for infrastructure investments and operations obligations.	With lower levels of community resilience, there is a greater reliance on public resources.	With a shrinking fiscal capacity, investments in environmental quality and EBA may be limited

resilience. For local governments, out-migration might be adaptive to the extent that people leaving high-risk sending zones are otherwise a drain on public resources. However, in the long term, the exodus of climigrants may be maladaptive if their fiscal capacity is undermined by a declining tax base. For adaptation policy makers and planners, understanding these trade-offs is key for ensuring distributionally and procedurally fair outcomes.[69] This is not only crucial for advancing climate justice, but it is also necessary to advance the public legitimacy of investments in climate adaptation.

Climate Adaptation Policy and Planning

Climate adaptation policy and planning (APP) in the United States is a rapidly emerging area of specialized practice. While there have been several decades' worth of policy development in climate mitigation and renewable energy, the adaptation sector has been slow to catch up. This is the case not only for the development and deployment of APP but also for the development of the people and expertise necessary to support the practices.[70] The reason for this delayed maturation is often attributed to the fact that climate adaptation was a dirty word for a long time.[71] The general argument was that if people focused on adaptation, then they would do so at the cost of more focused efforts in climate mitigation. Why should one mitigate GHG emissions when one can just adapt to the impacts? This flawed argument was reinforced by early integrated assessment models (IAMs) largely designed by economists in an attempt to identify a range of climate policies that would find some balance between acceptable emissions and the lowest measure of GDP impacts associated with the costs of decarbonization and adaptation.[72] Decarbonize as much and as fast as possible never seemed to be the right answer.

The anti-adaptation bias was problematic, but it nonetheless defined the landscape for international climate policy up until the 2015 Paris Climate Accord, at which point adaptation was formally elevated to its rightful place in parity with climate mitigation in climate policy.[73] This chapter does not attempt to outline the historical evolution of APP in the United States. Rather, it uses this pivotal moment in history as a starting point for understanding a range of recent activities in the public sector. Chapter 4 goes on to explore private-sector adaptations that are increasingly memorialized in market signals that should not be overlooked, particularly by regulatory agencies. Although much of the public-sector activity is concentrated in state and local governments, the federal government, at one point, played an important role in APP.

Federal Adaptation Policy and Planning

Climate adaptation is a dirty word in the second Trump administration. However, understanding the historical development of federal APP is instructive of where federal, state, and local governments may carry forward the work in the future. The development of APP in the federal government has taken two distinct tracks that would later merge at the agency level. The first track is defined by engineering or disaster resilience. After 9/11, resilience became an important concept in national security, which reinforced an all-hazards framework that integrated human, technological, and environmental elements of risk and risk mitigation. Engineering and disaster resilience were viewed as a way of filling in the blanks when formal signal and human intelligence failed to support targeted risk-mitigation activities. Given the wide range of potential terrorist targets, resilience was as much an organizing metaphor as it was a clearly articulated policy agenda or design standard. After the failures of the federal government following Hurricane Katrina in 2005, resilience formally found a place in the White House National Security Council (NSC).[74] At this juncture, resilience was primarily positioned as an element of domestic national security within the domain of the protection of critical infrastructure systems.

By the early 2010s, a parallel track in adaptation was emergent at science-based agencies, such as the National Aeronautics and Space Administration (NASA), the National Oceanic and Atmospheric Administration (NOAA), the Department of the Interior (DOI), and the EPA. Their interest had arisen because climate impacts were already being observed, and these agencies were trying to figure out how they were going to shape their statutory missions. For instance, at the State Department, adaptation was playing an increasingly important role in climate diplomacy because major state actors, such as India, were demanding compensation for loss and damage and support for adaptation.[75]

At this juncture, agency pioneers such as NASA's Sam Higuchi began self-organizing groups such as the Interagency Forum on Climate Risks, Impacts and Adaptation.[76] This forum was not directed or paid for by the

White House. Rather, it was an ad hoc group of people from a wide range of agencies and departments, from the Department of Justice to the Federal Highway Administration, that either had some interest or ongoing work in APP. The forum would go on to host a who's who of American adaptation experts that would drive much of America's APP in the coming years thanks to the foresight of Higuchi, the DOI's Tom Fish, and Ann Kosmal of the General Service Administration (GSA), among others.

In some agencies, such as the EPA, resilience and adaptation were separated into two programs that were not in dialogue with each other. Administrators, such as the EPA's Gina McCarthy (2013–2017), came of age in the era before climate change impacts, and they often lacked an interest or background in resilience and adaptation. In addition, dual unintegrated resilience and adaptation programs were often a consequence of congressional appropriations. With Republicans often having an outsized influence on agency budgets during the past several decades, national security items, such as resilience, often got a free pass. More than once, congressional staffers noted that Republicans could relate to the concept of engineering or disaster resilience because it was an inherently conservative concept, but the transformative ideas behind adaptation were a step too far in ideological terms. As has been widely understood in Washington, Republicans have ignored the opportunities for APP at their own peril, particularly given the disproportionate risks from climate impacts facing the Sun Belt.

The Obama administration used the occasion of Hurricane Sandy in 2012 to bridge resilience and adaptation programs across the federal government. With the political and financial support of the Rockefeller Foundation, this policy experiment was advanced under the framing of resilience, much to the consternation of the federal adaptation contingent. A senior NSC official noted that this friction led to unusually "heated" debates in the White House, with scientists arguing that the hazard-mitigation perspectives of resilience in national security were entirely inadequate to handle the complex socio-ecological and system dynamics at work with climate change. They argued that this Americanization of resilience would also put the United States out of step with the international community, which was moving at full speed with developments in adaptation science. Ultimately, the Obama administration sided with the national security framing, in part, because they thought that future congressional appropriations would be much easier to get through under a resilience framing.

Under the leadership of the resilience pioneer Alice Hill[77] and others, the Obama administration led a wide range of policy efforts, including the incorporation of resilience into the "White House Climate Action Plan" and the promulgation of key directives.[78] This included empaneling an interagency unit called the Community Resilience Panel for Buildings and Infrastructure Systems that built a national network of federal, state, and local stakeholders engaged in engineering and community resilience planning.[79] Among various policies, Presidential Policy Directive: Critical Infrastructure Security and Resilience (PPD-21) was uniquely impactful, with a well-developed outline for a formal set of interagency responsibilities driving the coordination of engineering resilience in infrastructure systems.[80] Although the first Trump administration undid PPD-21, it found a new life in revised form in the Biden administration under the auspices of national security. Unfortunately, all signs point to a broader dismantling of the federal government's role in engineering resilience for critical infrastructure systems in favor of greater state and local responsibility under the second Trump administration's Executive Order 14239, titled Achieving Efficiency Through State and Local Preparedness.

Much of the federal APPs that the second Trump administration is dismantling were created during the Obama administration. A review of more than 100 resilience policies and plans during the Obama administration found a strong focus on the shocks of natural disasters, but it also found that almost half of the policies and plans contained a meaningful incorporation of climate change and adaptation in some dimension of the work.[81] Despite a focus on resilience, adaptation found its place at an agency level with the publication of adaptation plans by agencies and departments.[82] However, these plans were more or less supplements to existing sustainability reporting requirements, and they were largely inward looking, with a focus on how agencies could adapt their facilities and the delivery of their core regulatory and administrative services.

Up until the second Trump administration, these plans were coordinated through the Office of the Federal Chief Sustainability Officer within the Council on Environmental Quality (CEQ). Across previous Democratic and Republic administrations, resilience had the advantage of being within the high-profile NSC. Adaptation had the disadvantage of being tucked deep into the recesses of a much lower-profile CEQ. Unfortunately, this office was never sufficiently staffed with adaptation experts and relied almost exclusively on federal employees who served on temporary details from other agencies. The resulting turnover of leadership limited the office's capacity to elevate the quality and depth of agency-level APP.

The first Trump administration was quick to disassociate its administration from the climate accomplishments of the Obama administration. This transition resulted in a great deal of inertia and the exodus of many experienced public servants. Although a reliance on agency details has its disadvantages, one advantage is that subject matter experts (SMEs) were retained in some key areas as protected civil servants. While APP largely came to a standstill during the first Trump administration, there was one bright spot that endured over the years. The Department of Defense (DOD) continued to proclaim that climate change was a risk to national security,[83] and it continued to support a wide range of APP and resilience activities, for a wide variety of areas, from base planning and logistics to force deployment.[84]

To the surprise of many experts, the subsequent Biden administration was slow to act on climate adaptation. Its core efforts were initially led by key personnel details from the GSA and DOD who had maintained mission-critical activities through the first Trump administration. They inherited a system that was still defined by two tracks of resilience and adaptation across the agencies. At the time of the transition, agencies severely lacked not only APP SMEs, but they also lacked expertise in sustainability and climate mitigation across the board. Multiple sources from inside the Biden White House tell a similar story. Early on in the administration, it was decided that they would go after major climate mitigation goals while they had the legislative opportunity to do so. APP would take a second seat for two fundamental reasons. First, the administration would need all the people that it could muster to roll out what would be the massive 2022 Inflation Reduction Act (IRA). With limited administrative capacity and hundreds of billions of dollars to spend, they had little bandwidth for APP.

Second, as a White House insider noted, it was determined that APP was "bad politics" for the Biden administration. Adaptation is often local. It often involves tough trade-offs. It can be a messy business, and the administration did not want to be seen as the arbiter of the winners and losers associated with adaptation trade-offs. Outside of select tribal communities, it surely did not want to be associated with helping facilitate the managed relocation of communities, as some progressives might rhetorically argue that this "forced displacement" was further marginalizing communities. At the same time, the administration felt obligated to follow through on its pledge to help frontline communities facing a range of environmental injustices. Their major policy accomplishment in this regard was the implementation of Justice40, which required that 40% of benefits from climate and energy investments go to disadvantaged and marginalized communities.[85]

Consistent with Justice40, the progressive politics of the time dictated that environmental justice (EJ), climate justice (CJ), and Indigenous people's justice drove the central focus of the administration's resources applied to APP. This sounded perfectly noble, but it ran headlong into some problems. First, EJ communities have had—and continue to have—major issues with toxic pollution that require immediate attention, such as access to clean water. While long-term strategic investments associated with APP were on their radar, EJ advocates often had different short-term priorities. Second, many of the most vocal EJ and CJ advocates who would normally staff an administration had very little experience in the techno-managerial aspects that drive APP practice. Despite the presence of many seasoned EJ professionals, their expertise was largely, but not exclusively, focused on a specialized area of environmental advocacy, litigation, and regulation. Many of those seasoned EJ experts with significant experience in APP were already spread thin with existing work and on-the-ground crises.

Many EJ advocates who did make it into the Biden administration did not last long once they hit legal, political, and procedural walls.[86] They quickly realized that the political rhetoric of the administration had hard limits, especially in Congress and in the courts, which were both skeptical of policies such as racial preferences for EJ communities.[87] Despite the rhetoric, research has suggested that Republican governors do not seem to mind helping disadvantaged and marginalized people in a manner not too different from Justice40 when they are spending federal climate investments.[88] It is a different story when they are spending their own money.

The Biden administration misunderstood this misalignment between rhetoric and experience, and the result was a frustrating policy inertia often shaped by undefined progressive jargon. This lack of a clear policy direction frequently alienated senior agency personnel who were tasked with building programs that lacked a realistic endgame. Building an "equitable world" is not the kind of concrete goal that federal bureaucracy is used to addressing. This often led to friction among agency personnel and appointees. As one senior agency adaptation leader noted, "I rather have a clear adversary than to have constant friendly fire."

To make matters more challenging, the administration ultimately lost ground with a lack of congressional budget support for adequately staffing the government with the people necessary to do the work. Between understaffing and a lack of a clear strategic direction, the turnover in APP leadership was a barrier to sustained policy development. In the competition between climate mitigation and adaptation, mitigation was the overwhelming winner within the Biden administration. This makes sense. Climate mitigation has long been considered the priority in international policy, with adaptation viewed as filling the gaps of incomplete decarbonization measures.

It is worth noting that any analysis of adaptation interventions outlined in the previous section of this chapter should include an evaluation of the GHG impacts of adaptation decisions—and vice versa. Building vast amounts of concrete coastal infrastructure, increasing the application of fertilizer to maintain crop yields, and pumping water farther distances are all forms of adaptation that have significant GHG emissions. At the same time, climate mitigation can also limit adaptive capacity. Converting corn and sugar crops into biofuels drove up food and cereal prices, and the resulting cost burden undermined the adaptive capacity of low-income households around the world.

Closer to home, energy utilities always want to pass along the costs of stranded fossil-fuel-generation assets and adaptation investments. This often means that, for some, utility prices are going to go higher before they go lower. This kind of inflationary economic stress can limit household adaptive capacity. It might sound like a policy abstraction, but plenty of households make trade-offs between food and air conditioning in the summer. As the EIA found, "Nearly one-third of U.S. households (31%) reported facing a challenge in paying energy bills or sustaining adequate heating and cooling in their home."[89] In the future, people and firms might very well just move to where the electricity from renewable energy is the cheapest. In this sense, what feels like mitigation-driven maladaptation in the short term is really setting the stage for transformative adaptation in the long term. To their credit, these were the types of detailed long-term policy trade-offs that the Biden administration was deeply invested in addressing.

In the near term, climate-attributed disasters have been accelerating in frequency and intensity. As with all disaster trend analysis, some of the widening expanse of climate risk is attributable to growing measures of exposure, as the population expands into higher-risk geographies. The trends are undeniable. From 1980 to 2024, the United States averaged about $64 billion in inflation-adjusted (2024) losses a year.[90] In the 1980s, the figure was just under $22 billion and went up to $33 billion in the 1990s. In the 2000s, annual losses doubled to just under $62 billion. In the 2010s, with storms such as Hurricane Katrina, losses averaged just under $100 billion a year. From 2019 to 2023, the figure went up to $123 billion, with just the last three years (2022–2024) averaging almost $153 billion in losses a year. Not all extreme events are attributed to climate change, but event-attribution science suggests that upward of nearly 50% of economics losses may be attributed to climate change.[91]

The survey of federal APPs—at least before the second Trump administration—starts with the frontline policies and practices associated with disaster response and recovery. As represented in Figure 3.3, FEMA is just 1 of more than 30 federal agencies and departments that have functional responsibility for supporting post-disaster activities across various sectors from housing to environmental resources.[92] Although the second Trump administration has positioned FEMA as a dysfunctional and ineffective agency, it is worth noting that FEMA has been grossly understaffed by thousands of workers.[93] At the height of hurricane season in 2023, FEMA was so under-resourced that it actually ran out of money.[94]

This might be a generous defense of an agency that struggled to keep up with a changing climate in the years before Trump's second term. For instance, the governing policies known as the "National Response Framework" and the "National Disaster Recovery Framework" had just four references to climate change and one reference to climate adaptation between the two documents combined.[95] This is problematic because decisions made during post-disaster recovery set up the path dependencies for future adaptation options. Indeed, the long-standing FEMA policy of strictly limiting post-disaster public assistance grants to rebuilding damaged infrastructure in its exact pre-disaster condition undermined the capacity of these systems to later incorporate engineering resilience and adaptation features.

Congress previously amended the governing Stafford Act[96] several times over the years to create programs that serve to support APP. In 2018, Congress required a mandatory 6% set-aside from estimated post-disaster expenses to fund a program known as the Building Resilience Infrastructure and Communities (BRIC) program.[97] The program was designed to offer

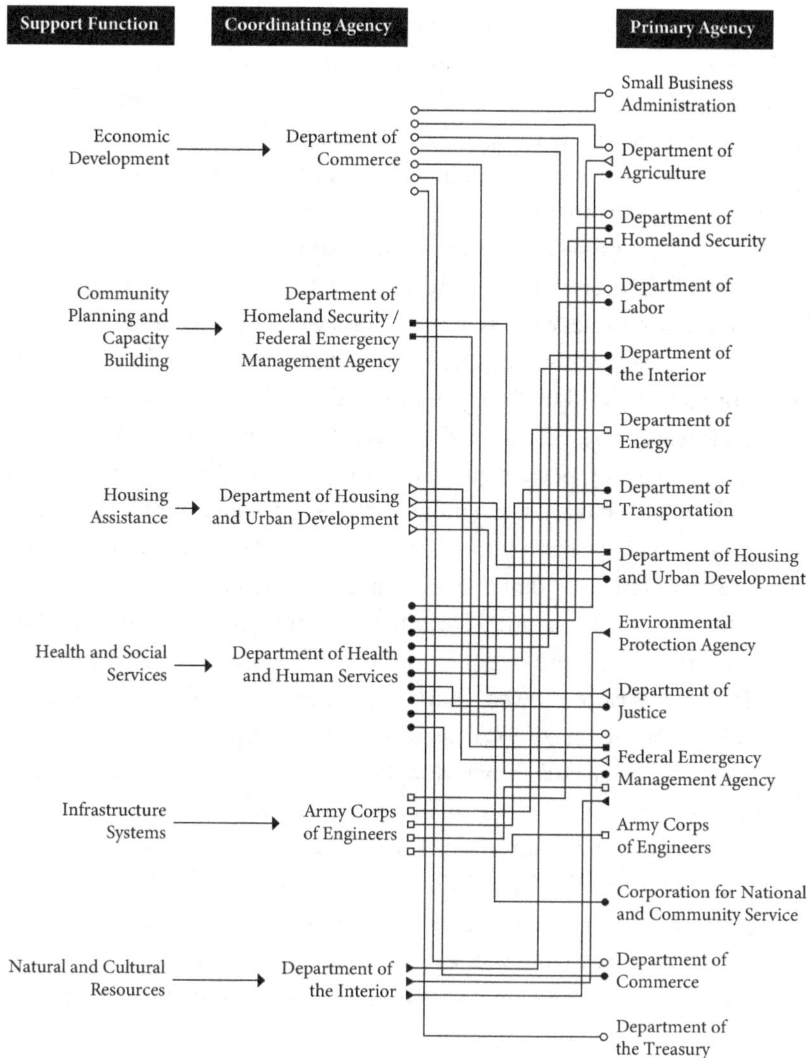

Figure 3.3 Post-Disaster Response and Recovery Functions in the U.S. Federal Government Before the Second Trump Administration. Design Credit: Oliver Oglesby, Alper Turan, and Jesse M. Keenan. *Data Source*: Government Accountability Office (GAO). 2023, May 17th. "FEMA: Opportunity to strengthen management and address increasing challenges." GAO-23-106840. Washington, D.C.: U.S. Congress

grants to fund hazard mitigation and engineering resilience investments not only for states but also for local governments and tribal nations. It would later become the main conduit for IRA and Biden-era dollars in resilience, including investments in things like stronger building codes that include engineering resilience requirements and nature-based solutions (NBS)—a form of EBA. Unfortunately, owing to this climate connection, BRIC was frozen in place by the second Trump administration.

Prior to BRIC, there was already a Hazard Mitigation Grant Program (HMGP) in place to steer funds mostly to states and was designed for more conventional hazard mitigation and disaster preparedness investments. The BRIC program was designed to give innovative local jurisdictions a shot at funding. However, the HMGP had evolved to support a great deal of APP. Over time, upward of 39 state governments used the HMGP as a means to integrate their state hazard mitigation plans (SHMPs) with their state adaptation plans.[98] This was stimulated by 2016 and later 2023 guidance from FEMA that explicitly said that climate change should be included within the scope of SHMP development.[99] Despite this momentum, Congress and the Biden administration were intent on doubling down on the BRIC program.

The BRIC program allowed FEMA to potentially bypass state bureaucracy and politics to make local investments in a much broader realm of engineering resilience and climate adaptation. However, the program struggled to get money out the door, as many local communities either did not have the staff capacity to administer BRIC or struggled to raise the required 25% local cost share.[100] This reinforced the underlying problem that federal pre-disaster and post-disaster benefits have disproportionately gone to wealthy communities that have the resources to plan for and administer federal grant programs.[101] Although BRIC included provisions for small and under-served jurisdictions to pay as little as a 10% local cost share, this cost burden was beyond the capacity of many, if not most, small towns. In 2022, Congress passed legislation that required FEMA to identify and map Community Disaster Resilience Zones (CDRZs).[102] While these high-risk and under-served CDRZs are eligible for the 10% reduced cost share, the currently frozen BRIC program might eventually need to be modified to a 0% cost share to be effective. As a consequence of a lack of capacity and matching resources, hundreds of millions of BRIC funds went unspent and projects stalled[103]—this was sometimes referred to as hitting the "BRIC wall."[104]

Perhaps one of the most relevant categories of APPs relates to the federally funded buyouts of properties in high-risk areas, including flood zones.

While FEMA and the Department of Housing and Urban Development (HUD) have the most high-profile programs, the Department of Agriculture (USDA), NOAA, the National Park Service, and the U.S. Army Corps of Engineers (USACE) each have the regulatory capacity to buy out properties.[105] The USACE has taken the position that either voluntary buyouts or eminent domain actions advanced by local government partners are likely going to be required to fully execute coastal climate adaptation and resilience projects in the future.[106] This position has raised the alarm among some observers that compulsory managed relocation may not be too far off into the future.[107]

Through the FEMA and HUD programs, local and state governments may apply for grants to buy out local property owners to reduce the risk of future losses. For instance, homes that are in repetitive loss areas that have been regularly flooded in recent years may be given priority. This not only reduces private losses, but it also reduces future public losses associated with infrastructure impairment, as well as losses to the federal government from NFIP claims. From 1989 to 2017, the FEMA program engaged in 43,633 voluntary buyouts in 49 states.[108]

Adaptation scholars such as A. R. Siders and others have focused on buyouts as an active area of research because many commentators believe that the United States will be forced to engage in more active voluntary buyouts as a first step towards compulsory managed relocation in the future.[109] From one point of view, these programs tend to benefit wealthier homeowners in a manner that amplifies socioeconomic inequality to the extent that the money could have been utilized for more targeted alternative adaptation interventions for more vulnerable populations (e.g., low-income renters).[110] While buying out properties has been estimated to reduce risk,[111] there are still outstanding questions about whether the participating properties should be left vacant or rewilded, or whether there are—potentially problematic—opportunities to rebuild "resilient" housing on the land.[112]

This post-buyout conflict highlights three fundamental problems. First, participating households often cannot afford to relocate within the same community or jurisdiction. Research has found that people who relocate often move to an area with a higher poverty rate and many simply move to another high-risk flood zone.[113] Second, non-participating renters are often left without any transition support, just as local rents go up as the local housing stock declines.[114] Third, the overall scarcity of land has the potential to drive inflationary pressures that reverberate across the housing stock. To partially address these problems, some have argued that the local government can either support the redevelopment of housing that is adapted to flood risks (e.g., elevated), or they can rent the homes or offer life estates to participating households who are able to retain possession of their properties for the balance of their natural lives.[115] Of course, allowing people to remain in high-risk areas likely just delays the pill that eventually has to be swallowed.

The most promising area of policy development relates to the idea of "buy-ins," where local governments strategically develop a local receiving zone to retain the local population and taxbase.[116] It is yet to be determined the extent to which local governments have the development appetite or the available land necessary to build housing commensurate with the scale of future buyouts or buy-ins. Even with subsidies, unit costs may not be commensurate with the price points that are affordable for participants. To FEMA's credit, they previously adopted guidance that dictates that applicants who intend to advance relocation projects should have a plan in place for where those people are going to go.[117] FEMA wanted applicants to start planning for local receiving zones.

Whether it is managed relocation, community relocation, or simply planned decommissioning, this was an active area of APP development prior to the second Trump administration.[118] The term "planned relocation" even made into *NCA5* in 2023, which defines it as "a form of mobility in response to direct climate impacts and/or indirect economic costs of estimated and projected climate impacts. Planned relocation, also referred to as managed retreat, is typically initiated, supervised, and/or implemented by public, private, and civic stakeholders and involves small communities and individual assets but may also involve large populations."[119] Although HUD even went so far as to release an "implementation guide" on community-driven relocation, there was limited evidence of actual implementation.[120]

Jointly chaired by FEMA and the DOI, the Biden administration launched the Interagency Community-Driven Relocation Subcommittee in 2022 with the ambition to develop a whole-of-government approach to manage relocation. Whether it is coordination with the Small Business Administration for the transfer of business registrations or coordination with the EPA for endangered species evaluation in receiving zones, the cross-jurisdictional issues were immense. At a fundamental level, the subcommittee wanted to figure out "what should the prerequisites be for communities to be included [in a planned relocation program]?"[121] In a July 2023 meeting, CEQ raised the question of whether vulnerability assessment studies should be undertaken to qualify participating communities in the future. Kelly Main, from the nonprofit Buy-In Community,[122] noted with a measure of exasperation, "Skip the risk assessment. There are plenty of people who are ready to move *today*."

CEQ was not naive about the challenge. Rather, they were simply starting from scratch. Inspired by a call from the Government Accountability Office (GAO) for a pilot relocation program,[123] the IRA provided $115 million to support the relocation of 11 tribal nations.[124] By the end of the Biden administration, the work of the subcommittee had not materialized into anything substantive beyond a final report, and this left tribal relocation advocates like Robin Bronen frustrated by a lack of progress. Interviews consistently reported that the execution of the pilot program was not sufficiently staffed or funded at a level necessary to actually relocate the selected tribal communities. Interviewees working with the subcommittee speculated that this task was likely bigger than the one project manager at FEMA could possibly manage. It seemed unlikely that a single—albeit well-qualified—project manager had the capacity necessary to steer dozens of agencies in a coordinated effort to move people at the scale contemplated by the program. As one interviewee noted, "A thousand civil servants might not be able to make a dent in this work."

Beyond the domain of FEMA, the Biden administration picked up where the Obama administration had left off with APP across federal departments, agencies, and administrative units led by the Office of the Federal Chief Sustainability Officer at CEQ.[125] As outlined in Figure 3.4, a vast array of federal offices and agencies were tasked with adaptation policy, planning, and research during the Biden administration. To orchestrate these efforts, roughly two people at any given time in the White House were in charge of APP activities. This often came as a surprise to people who follow climate change in the United States. By some accounts, DOD alone had hundreds of people working in APP and resilience, and the Department of Energy (DOE) had thousands who worked on climate mitigation in some form or fashion. Despite the limited staff, the Biden administration was able to successfully coordinate the development of adaptation plans from 28 departments and agencies by 2024.[126]

In 2025, the second Trump administration brought all this work to a grinding halt. While the full coverage of Biden-era adaptation plans is beyond the capacity of this chapter, there were a number of key activities

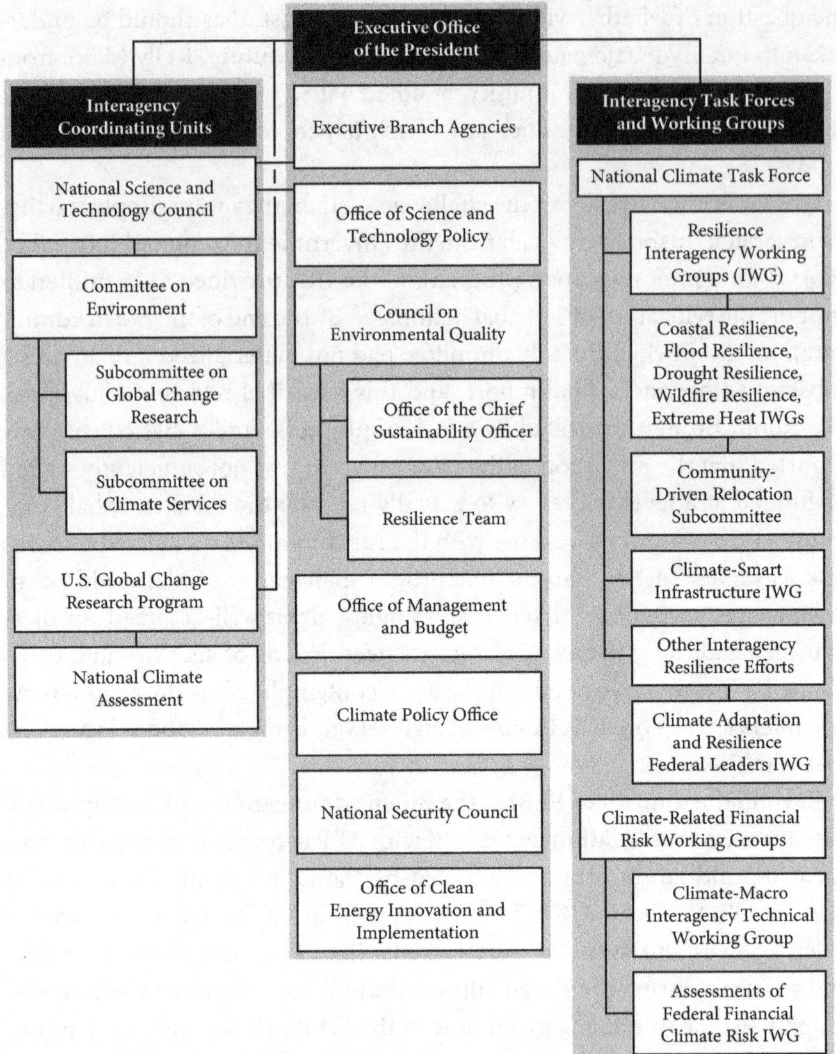

Figure 3.4 Organization of Climate Adaptation Policy, Planning, and Research Within the Executive Branch Under the Biden Administration. Design Credit: Oliver Oglesby and Jesse M. Keenan. *Data Source*: Government Accountability Office (GAO). 2024, August. "Climate Resilience: Congressional action needed to enhance climate economics information and to limit federal fiscal exposure." GAO-24-106937. Washington, D.C.: U.S. Congress

that speak to the diversity of challenges that endure no matter who sits in the White House. Indeed, federal APP innovations may need to be replicated in some form by states as they attempt to fill gaps left by declining federal involvement in pre- and post-disaster policies and programs.

Prior to the second Trump administration, the adaptation plans of the agencies shared in a few things common. First, they shared the ambition to apply climate risk management practices to the management of their assets. This required that agencies train their personnel in relevant climate science and adaptation practices to increase not just their literacy but also their institutional adaptive capacity for making climate informed decisions in the future. As was previously referenced, a lack of capacity in state and local governments is considered a significant barrier to supporting APP. Second, the agency planners were increasingly conscious of how their decisions might negatively impact EJ communities. This was a direct reflection of Biden-era executive oversight that required a broad evaluation of trade-offs among stakeholders. This reflexive point of view likely would have had internal co-benefits in the long run as it provided a mechanism for closer engagement with constituents.

Third, the agency plans benefited from collaboration across agencies. These plans offered valuable insight into how APP development operates as a social process. Adaptation pioneers like Sam Higuchi and Tom Fish understood this well before anyone else. The problems that these agencies face are complex, and the most effective solutions are often found in cooperative development. This is almost never successfully driven by a presidential administration. Rather, it is driven by happy hour exchanges at forums among like-minded professionals who are deeply engaged in these issues. They share ideas, research, data sets, contacts, and they indirectly put pressure on everyone from the Office of Management and Budget (OMB) to the Federal Accounting Standards Advisory Board to go a step further. Some agencies, such as the Department of Homeland Security (DHS), DOD, and the Department of Transportation (DOT), were extremely well socialized across the ecosystem of adaptation professionals. Other agencies such as EPA and HUD were not as well connected, and this lack of connection was reflected in their adaptation plans.

Upon closer inspection, the possible implications for APP across these agencies were vast. The USDA was planning to shift crop production requirements that would have had significant implications for migrant labor and the housing for farm laborers, as different regions focus on different crops.[127] As such, they sought to make investments that advanced everything from community resilience to the adaptation of housing in rural communities. For people living in the WUI in the West and Mountain West, USDA sought to remove dead trees, monitor invasive species, and manage the forests in order to manage devastating wildfires.

In terms of the environment, the DOI was working hard on adapting the management of federal lands and the nation's natural resources.[128] Some agencies were not only planning for climigration, but they were also planning for the shifting ranges of flora and fauna. NOAA was thinking about how to support fishing communities whose economic base was changing, often because their catch were slowly moving north out of range to deeper and colder waters.[129] Through adaptation investments in things like different nets to catch different fish, NOAA was working on supporting a broader transformation of America's coastal fisheries.

DOD was working on everything from APPs for communities and supply chains that support the defense industrial base to the integration of science and intelligence to better understand emerging climate hazards.[130] NASA was playing a similar role in the translation of climate and impact science to support decision making and portfolio management.[131] This was in addition to their APP activity at high-risk coastal launch and research facilities. Within DOD, the USACE was active in researching regional adaptation infrastructure projects that might be needed from coast to coast.[132] They were developing new indicators for adaptation and were working with other agencies to publish the best available projections for SLR. They were also active in providing technical assistance to state and local APP activities thanks to funding from the IRA.

DHS was working on resilience standards and performance measures within its workforce, supply chain, and at its facilities.[133] They sought to integrate adaptation and resilience consideration into critical infrastructure assessment and intelligence systems, including cyber-physical systems. They even took steps to develop better intelligence and planning around climigration and mass migration associated with climate change. With FEMA under their authority, DHS had its hands full with APP development and execution. FEMA's major accomplishment had been the publishing the of final regulations for the Federal Flood Risk Management Standard (FFRMS).[134] Initiated by the Obama administration, canceled by the first Trump administration, and completed by the Biden administration, the FFRMS set the standards for when and how buildings and infrastructure using federal financial resources may be placed in high-risk flood and SLR zones. The second Trump administration quickly revoked implementation of the FFRMS.

In terms of the built environment, the EPA[135] had integrated climate adaptation considerations into a range of infrastructure programs from brownfields grants to state revolving loan funds for water infrastructure.[136] NOAA was actively working with the National Institute of Science and Technology (NIST) to incorporate engineering resilience and community resilience standards within the planning, design, and engineering of critical infrastructure.[137] HUD was integrating APP into their buyout, building code, and block grant programs.[138] They even launched a new program, known as the Resilience and Energy Assistant Loan (REAL) program, to support the retrofit of properties with engineering resilience investments. They were also keenly studying the implications for lending and underwriting of mortgages in high-risk areas under their supervisory activities at the Federal Housing Finance Agency (FHFA), and they were supporting resilience investments made by public housing authorities.

The GSA—which manages the largest real estate portfolio in the world—was actively assessing climate vulnerability in supply chains, utilities, communication systems, and at historical and cultural assets.[139] They were even incorporating data about the elevation of assets into their portfolio management system to help manage flood risk. They were actively evaluating climate risk in nearly every material aspect of asset management, and they incorporated these analyses into National Environmental Policy Act (NEPA) evaluations. They were also engaged in developing disclosures made pursuant to guidance promulgated by the internationally recognized Task Force on Climate-Related Financial Disclosure (TCFD), as will be further contextualized by private-sector activities in chapter 4.

Although the DOE was primarily focused on climate mitigation, they had an active adaptation portfolio.[140] This included advancing engineering resilience standards and investments in energy infrastructure. DOE's standard-setting activities covered everything from increasing the wind and heat loads of transmission lines to simple things like insulating pipes and conduits that might freeze and break in an extreme cold event. They were making adaptation investments in hydropower to accommodate shifting precipitation patterns. Like the USDA, they were actively investing in land management practices to reduce wildfire risks.

In terms of transportation infrastructure, the DOT was working on addressing everything from reduced access to shopping docks and shoreline navigation to the increased operations and maintenance (O&M) costs from increased extreme heat and ice impacts on roads.[141] DOT had been active in requiring that adaptation and resilience considerations be incorporated into conventional grant and loan application processes. This highlights that for most agencies, there was no such thing as a stand-alone adaptation project. Rather, the goal was to mainstream climate considerations into everyday investments.

Finally, the Social Security Administration (SSA) was even thinking about where future SSA offices would need to be relocated.[142] This meant that they needed to "maintain plans for possible relocation in the case of inundation or large damages."[143] In the future, the SSA will need to have offices to support local populations. If some portion of the population moves off the coasts and out of drought-stricken areas, then their offices are likely to follow.

Overall, federal agencies advanced a great deal impactful and effective APP, but many internal observers struggled with understanding how these efforts connected to challenges that often crossed agency jurisdictions. In 2023, the Biden administration released the "National Climate Resilience Framework" as a first step toward developing a national adaptation strategy.[144] At least this was the ambition. Much to the consternation of staff, they were given only a handful of months to develop the framework. The final product was a notable collection of APP accomplishments, but it lacked a normative vision for a national strategy that could drive APP across jurisdictions, sectors, and geographies. It was not a national plan. It was a set of accomplishments edited in a style that was intended to appeal to party politics. Journalists from major media sources were not inclined to cover its release because, as one journalist noted, there were not any new policies to cover.

This was a missed opportunity. The United States needed a National Adaptation Plan (NAP). According to the United Nations Framework Convention on Climate Change (UNFCC), the 195 signatories to the 2015 Paris Accord agreed in Article 7 that "[a]ll [p]arties should engage in adaptation, including by formulating and implementing National Adaptation Plans, and should submit and periodically update an adaptation communication describing their priorities, needs, plans and actions."[145] Not only had the United States not complied with this component of the Paris Accord; it was the only G20 country that had not even started the process of developing an NAP. While NAPs were initially intended to be a path for the fair and negotiated distribution of adaptation financial assistance for developing and least developed countries, all participating countries would later commit to developing an NAP. From Japan to the Russian Federation, major countries had produced NAPs. Diplomats at the State Department were frustrated by the Biden White House's lack of focus for recognizing the opportunities and obligations associated with the NAP development process. Thanks to hard-working career diplomats, the Unites States eventually submitted an official NAP with the UNFCC just days before the launch of the second Trump administration.[146] Unfortunately, the submission was rushed to completion in a matter of weeks with the full recognition that it should have been developed over the course of years.

The NAP development process represented an opportunity to develop a unified strategic approach to complex problems that cut across conventional jurisdictional boundaries. In more practical terms, interviewees at the UNFCC consistently cited the proposition that private-sector adaptation could be more effectively guided, if not steered, through a well-developed NAP. They argued that it could send important signals about future policies and priorities. While some in the Biden administration questioned the ability of a NAP to steer an economy as large as the United States's, others argued that an NAP would be politically untenable.

Any plan that projected a normative vision for how the United States could adapt would have to wrestle with a determination of the winners and losers that would arise from the allocation of limited adaptation resources. Even for Democrats, it would be an untenable political position to project a sense of shared sacrifice. No administration wants to project a future with climigration or compulsory planned relocation. Everyone recognizes that the landscape for the "losers" of adaptation (e.g., non-wealthy climigrants or immobile populations in sending zones) is messy.

As the Biden administration cited, "Government action that targets constraints and market failures that impede adaptation should be most effective in supporting and enabling private action rather than crowding out actions that would have occurred anyway."[147] In other words, you are on your own, America. It will be the private sector and not the federal government that drives adaptation. The second Trump administration's dismantling of climate programs reinforces this hard truth. This book takes the position that while some people will bear the tremendous pain and sacrifice associated with climigration and planned relocation, many more people may the yield the benefits of being a "winner" in the pursuit a more sustainable American way of life. Unfortunately, the challenge of who gets left behind remains entirely undefined in federal policy.

State and Local Adaptation Policy and Planning

State and local governments are ultimately responsible for the vast majority of public-sector adaptation investments that are made and will be made in the United States. While the federal government made significant achievements in APP, these accomplishments were largely limited by Congress's abdication of the responsibility for resourcing adaptation. With the fiscal austerity and chaos of the second Trump administration, states are entirely on their own.

Public-sector adaptation requires significant investments, and the flow of funds is largely local even with a climate-friendly federal administration. In any given year, state and local governments make at least three-quarters of all infrastructure investments.[148] Being on the hook for public-sector adaptation is not necessarily a bad thing for local governments. After all, they are the ones who are best positioned to navigate local politics and to shape the participatory and democratic processes that will determine what (and who) should be protected and what (and who) will be left behind. Many local governments would rather make those decisions than have the federal government tell them what to do, particularly when it comes to making the tough decisions to relocate infrastructure and shift land uses away from high-risk areas.

State and local governments have several common objectives in adaptation planning. First, they are seeking to undertake vulnerability assessments to figure out the exposure and sensitivity of their populations and assets to projected climate impacts. In this sense, the state bureaucracy itself is the primary stakeholder.[149] They often hire large engineering conglomerates to do this kind of work. In big cities, these firms often underprice their climate vulnerability assessment services as a loss leader in the hopes of getting more sustained work in the future. These engineering firms are reasonably good at translating physical risk parameters from design events into the specifications necessary for everything from planning and design to operations and management of infrastructure systems. They are less good at dealing with the nuances of ensemble climate modeling and integrated assessment modeling. As such, their work is frequently challenged by local scientific experts as being riddled with uncertainties to the point that their work products are indefensible.[150]

Professional service firms often respond by noting that they are not being alarmist, but rather, they are advancing under the precautionary principle to plan for the worst. Unfortunately, given the ever-growing tail risks of climate change, identifying the worst-case scenario is extremely difficult, particularly in the context of indirect and systemic local infrastructural vulnerabilities. The challenge is that when one is planning for close to a worst-case scenario (e.g., the now outdated RCP 8.5) it may lead to a gross overestimation of risk that has two major implications. First, this overestimation may lead to the excess expenditure of limited public resources. These are not just marginal costs; this can sometimes be the difference between go and no-go in terms of project feasibility. There are also opportunity costs for the overestimation of risk in the face of deep uncertainty.

Second, major capital investments that plan for what officials think is the near worst-case scenario may also create significant path dependencies that limit future adaptation options. For example, Miami Beach made the decision to elevate many of its public roads.[151] The problem is that once they elevated one road, they found that they had to elevate other roads, as well as buildings and infrastructure, such as pumping stations, which bore the burden of faster-moving stormwater from steeper slopes on the elevated roads. It set in motion an almost never-ending process of elevating infrastructure that narrowed the path for future adaptation options due to its cost-intensive nature. Finding a balance between economic efficiency and the precautionary principle is not easy.

The second objective that state and local governments have in adaptation planning is the application of climate risk management tools, techniques, and principles to manage their services, facilities, and assets. If the first step is to figure out what the problem is, then the next step is doing something about it. The primary challenge that many local governments run into is that they often do not know what assets and facilities they have. One would think that governments have a central inventory of their assets uploaded into an integrated asset management database. As was previously referenced, the GSA did develop something like this. They even linked their asset database with flood risk data. Local governments almost never have anything like this. Their data is often a mix of paper and digital files that are strewn across departments. Climate tech firms like Forerunner have created a business model around simply bridging the analog and digital divide in flood and building data through the development of integrated asset-centered platforms.[152] While local governments might have a pretty good idea of what depreciable assets, such as fleet vehicles and heavy equipment, they own, they rarely have a sense of what real property they own. They often acquire a variety of properties—everything from tax foreclosures or gift transfers—and they rarely keep track of who owns what.

This legacy condition often requires that local governments spend a fair amount of time simply collecting information on what they do own. For those with the resources, they often use this exercise to assemble their inventory of geographic information systems (GIS) data files. Sometimes this is done during the first phase of a vulnerability assessment. Well-resourced local governments will go a step further to invest in updated data collection and integration, including investments in high-resolution LIDAR scanning. Once governments figure out what they have and where it is, the next body of work centers on organizing a system to triage these assets.[153]

At the top of this classification system, one might find "life-safety critical" assets, such as public hospitals. Assets such as public hospitals and water treatment plants are going to be prioritized for adaptation and resilience investments. At the bottom of the system are things that can be decommissioned and salvaged. A missing piece of this process often revolves around accounting. Once one knows that they own something, then they need to internalize it into their accounting. Once one knows that that an asset is or might be impaired, then they need to account for it, including in their operations and capital planning. There are strong disincentives for governments to properly account for assets because it means that they have to do something about their impairment. This creates a problem later on when local and state governments are opportunistically looking to make adaptation investments with federal money following a disaster. Without adequate planning, many governments are not prepared to hold their hands out to ask for federal support. As a consequence, local governments perpetuate an unsustainable status quo because all they can afford to do is to repair or rebuild lower quality infrastructure in high-risk places.

The final shared objective is to actually implement adaptation projects. In reality, aside from non-structural approaches to community resilience (e.g., supporting mutual aid nonprofits), free-standing adaptation and resilience projects are rare. It is much more common that local governments incorporate adaptation and resilience elements within conventional capital projects. This might be something simple like reinforcing the foundations of a bridge or installing larger stormwater pumps with their own backup power generation. Chapter 5 references a range of design and planning techniques in the built environment that incorporate resilience and adaptation under the broader umbrella of sustainable development.

As highlighted in table 3.2, local governments are tasked with implementing a wide range of ecosystem-based, technical, and institutional adaptation measures. Given the wide-ranging fiscal implications, local government implementation decisions are often positioned within a broader strategic allocation of limited resources.[154] Some governments may pursue a *no-regrets adaptation strategy*, which promises a material benefit even if

Table 3.2 Examples of Climate Adaptation Measures by Local Governments

Primary Impacts	Ecosystem-based adaptation (EBA)	Technical Adaptation	Institutional Adaptation
Extreme Precipitation and Flooding	Dual-purpose floodable landscapes	Excess pumping capacity and redundant power sources	Map expanded flood zones outside of FIRMs and restrict land use and zoning
Nuisance Flooding	Distributed drainage and water storage in public right-of-way	Elevation of roads and infrastructure	Change elevation requirements for new structures
Sea Level Rise	Support migrating wetlands and coastal habitats	Engineered adaptive capacity of coastal protection to handle greater loads	Up- and down-zoning to transition population, infrastructure, and tax base
Windstorms & Tropical Cyclones	Wetlands and reef restoration and conservation	Engineer critical infrastructure for high wind loads	Link hazard maps with land use and zoning performance requirements
Extreme Heat	Expand urban tree canopy	Increase albedo of roofs and public spaces	Expand public health communications and cooling center access
Drought	Transition to xeriscaping	Recycle wastewater and develop alternative reservoirs	Set efficiency standards and reprice water utility rates for scarcity
Wildfires	Restore natural hydrological flows and fuels management	Engineer for heat loading for critical infrastructure barriers	Modify building codes to reduce risk and limit development in WUI
Vector-Borne Disease	Clean waterways and remove invasive vegetation	Chemical vector control	Vector, clinical, and community-based surveillance

climate impacts fail to materialize as expected. For instance, a local government may develop a series of parks that can absorb and store excess stormwater. Even if they do not experience an extreme precipitation event with any regularity, the designs would likely allow them to right-size their pumps in a manner that reduced upfront capital and long-term O&M costs. In the worst—or best—case scenario without any floods, they still benefit from having built parks.

In other cases, local governments may pursue a "flexible adaptation strategy" that does not impose significant path dependencies. For instance, instead of building a massive concrete flood barrier that can accommodate a rapid SLR scenario, they may choose to deploy low-cost inflatable barriers. This may also be viewed as a *reduced decision-horizon strategy*, wherein adaptation investments are made that allow local governments to extend the time horizon for making decisions about larger, long-term structural investments. In this sense, non-structural approaches can buy time. Although this may be convenient for building political will while one waits for clearer signals about the nature of the risks, there will always be unresolvable uncertainties in the future. If one waits too long, then one runs the risk of running out of options and money.

The implementation of adaptation projects often means that local governments have to draw lines. On one side of the line are people who benefit, and on the other side of the line are the people who do not. When it comes to public finance, exactions, and property taxes, those who pay and those who benefit are rarely the same people. This simple bifurcation of benefits and costs drives a good deal of local politics. To make matters more challenging, adaptation plans often fail to clearly identify goals, benefits, or indicators to measure success.[155] Sometimes, proclamations about avoided future losses are as good as it gets.

People are frequently averse to change. Sometimes they fear that adaptation infrastructure projects are a slippery slope for more sustained development activity that will drive gentrification. As will be explored in chapter 5, they might very well have a cause for concern. Others simply distrust the government's capacity to take on mother nature. The smallest portion of people are either uninformed or misinformed about what the risks are and who stands to benefit from adaptation investments. Climate misinformation is vast and should not be underestimated. In general, experience demonstrates that people are usually well informed and prepared to engage in the public process of planning for adaptation. A vast majority of people just want to have an opportunity to hear from someone who speaks on their terms (e.g., nontechnical) about adaptation strategies and projects.[156] The more challenging task is finding professionals and public officials who actually want to listen to what people have to say.[157]

From a macro-perspective, state and local governments have engaged in a fair amount of adaptation planning in a short amount of time over the past decade. However, by all accounts, their experience with project implementation is limited. Well-resourced jurisdictions tend to dominate the landscape for adaptation planning. For smaller, poorer, and often more rural jurisdictions, they are either operating in the dark or they are relying on technical assistance from state and federal agencies offered by people who are usually unfamiliar with the constraints of local conditions. Technical assistance from the federal government is often defined by a group of highly qualified scientists and engineers who are organized by an off-site project manager. This group almost always lacks professional representation in law, accounting, public administration, public finance, and urban planning. As a result, locals and "experts" sometimes waste a lot of time trying to find a common ground.[158]

Despite the limitations, technical assistance is often critical for helping state and local governments fill gaps in their own expertise. It offers the prospect of bringing a fresh pair of eyes to look at a problem, particularly through the use of scenario planning.[159] It also allows for a great deal of socialization to the extent that professional relationships often drive everything from the exchange of new ideas to informal referrals of the best people or firms for the contract work ahead. Very often, technical assistance helps local governments figure out what grants and/or loan programs might work for them to support future adaptation investments. In the absence of such assistance, many local governments simply do not where to start.[160]

Some state and local governments have worked extensively to develop APP. Across the country, 39 states have integrated hazard mitigation and adaptation plans,[161] and 33 states have free-standing climate action plans that include some element of adaptation.[162] While the IRA provided $5 billion to 45 states and metropolitan areas to support climate mitigation for 96% of the U.S. population, no such resources went to adaptation planning.[163] Instead, state governments formerly relied on a patchwork of federal grant programs across agencies that are often sector specific. Luckily, recent research suggests that the development of a mitigation plan often helps propel states to carry forward with an adaptation plan.[164]

Despite the lack of federal support or direction, 17 states have free-standing adaptation plans, with another 15 states having either begun the development of a state plan or have sector-specific plans.[165] These numbers are up from just 13 states a decade ago.[166] Unfortunately, research highlights that political ideology plays role in inhibiting adaptation planning, with Republican-led states far behind the pack.[167] The high-risk state of Texas does not have an adaptation plan. This sets up a scenario where the states with the greatest measure of projected domestic climigration are not planning for adaptation. This can only make the prospects of out-migration from sending zones more likely as states fail to support successful adaptation.

According to the most recent count by the U.S. Census Bureau, there are 90,837 government entities in the United States, many of which are small municipal and special-purpose entities.[168] No one has a precise count on local government adaptation plans and activities. Based on a review of state plans and publicly available information, it is likely that only a few thousand entities have either done a climate vulnerability assessment or a hazard mitigation assessment that included climate change. By the same token, it is likely that the total number jurisdictions with some form of plan that incorporates adaptation is under 500. This is likely double the number of participating jurisdiction from a decade ago.[169] Only 45 of the top 100 largest cities in the United States have completed an adaptation study.[170] Of these largest cities, only a handful have advanced regional adaptation plans at a metropolitan scale.[171] By one estimate, even among the most active cities, climate resilience and adaptation are not well integrated within existing planning processes, with fewer than 37% of land use plans incorporating climate change.[172] A select review of more than 500 local adaptation policies in high-risk jurisdictions in the United States also suggests that over 20% of the policies are simply grants and subsidies to households and firms, suggesting a strong reliance on private-sector adaptation.[173] This is a theme explored in more detail in chapter 4. How far behind is the United States? By comparison, 167 cities in the European Union have adaptation plans.[174] While lots of barriers are thwarting local adaptation planning, the biggest barrier is often a lack of leadership.[175]

The United States needs climate adaptation leaders, people who are going to be clear and transparent about what the process is and what the goals are. These leaders also need to be flexible as the goals change. They need to be willing to include a wide variety of people in setting those goals. Analysis of adaptation interventions must include an analysis of the trade-offs and potential unintended consequences so that people can determine for themselves who the winners and losers are. Analysis must acknowledge that human and ecosystem adaptation processes exist in the same space. They need to understand that there are often inherent conflicts between resilience and adaptation and that the risk of maladaptation is always on the horizon. To do all these things—and to do them well—requires a great deal of capacity. This means that everyone from professionals to next-door neighbors needs access to training, education, and information. The are many accessible adaptation tools. The public sector needs people to use them.

Adaptation is not easy. It is messy. Maladaptation driven by a lack of leadership and analytical discipline will no doubt define the landscape for many places. While APP had progressed at a significant pace prior to Trump's second term, a vast majority of the U.S. population is not covered by a comprehensive adaptation plan. The potential unevenness of public-sector-driven adaptation means that some places are going to move forward toward sustainability and other places are going to be left far behind. For instance, if Republicans in the Sun Belt do not plan for adaptation, their lack of foresight is likely to accelerate the out-migration of highly exposed and cost-burdened populations. In this sense, local belief systems and a lack of political will that stand in the way of public-sector adaptation are likely going to be important factors shaping future climigration.

4

Market Signals in Housing, Real Estate, and Financial Services

Markets are increasingly sending signals that climate change is reshaping the costs, values, and availability of goods and services in the economy. Sometimes the adaptation of markets is contemporaneously understood as a maladaptation by consumers who bear higher costs in the near term. But this narrow orientation often clouds a more complete story that is coming into focus. As Senator Elizabeth Warren noted, "according to [the author of this book,] . . . 'The private sector has always adapted—one either adapts to new markets, products, or services or they go out of business. But the current calculus is more than a function of market share. It is a function of whether there will be a market at all.'"[1]

This chapter focuses on how climate impacts shape the supply and demand of housing, real estate, infrastructure, insurance, and other financial services. For some households and firms, the increased costs, shifting values, and declining economic output may serve as a powerful push factor for relocation. To make matters more challenging, both households and firms are reliant on the public sector for the provision of infrastructure and services that are increasingly under fiscal pressure from climate impacts. Chapter 3 highlighted the limited capacity and uneven effort of the public sector to address these impacts. State and local governments are largely on their own, with nearly zero resources and guidance from the federal government. With the exception of the largest and wealthiest jurisdictions, this has left public adaptation investments in the hands of local governments, who largely lack the institutional capacity necessary to make sustained investments in adapting infrastructure and public services.

This leaves a large vacuum that the private sector is slowly filling. This chapter seeks to advance the proposition that the private sector is well ahead of the public sector in steering capital that has begun to internalize physical climate risk. For the most part, local governments are not warning people that they live in a high-risk area; it is their landlord, mortgage lender, or

North. Jesse M. Keenan, Oxford University Press. © Oxford University Press (2025).
DOI: 10.1093/9780197641644.003.0005

insurance agent who are sending increasingly coherent messages. Indeed, some state and local governments are incentivized to mute these private market signals. People tend to believe that democratic institutions will determine who bears the burdens and benefits of climate impacts through the allocation of targeted public investments. Rather, it is the private sector that is likely to have an outsized influence in shaping these tough political decisions in the future. For some fortunate beneficiaries, there may very well be planned and orderly in situ adaptation or relocation. For others, there will be forced relocation at the hand of the market.

Housing and Real Estate

For many people, their homes and bank accounts are on the frontlines of climate change. As represented in Figure 4.1, these consumers are a small part of a vast architecture of convergent processes and sectors in the U.S. real estate industry that makes up roughly a fifth of the economy. Some

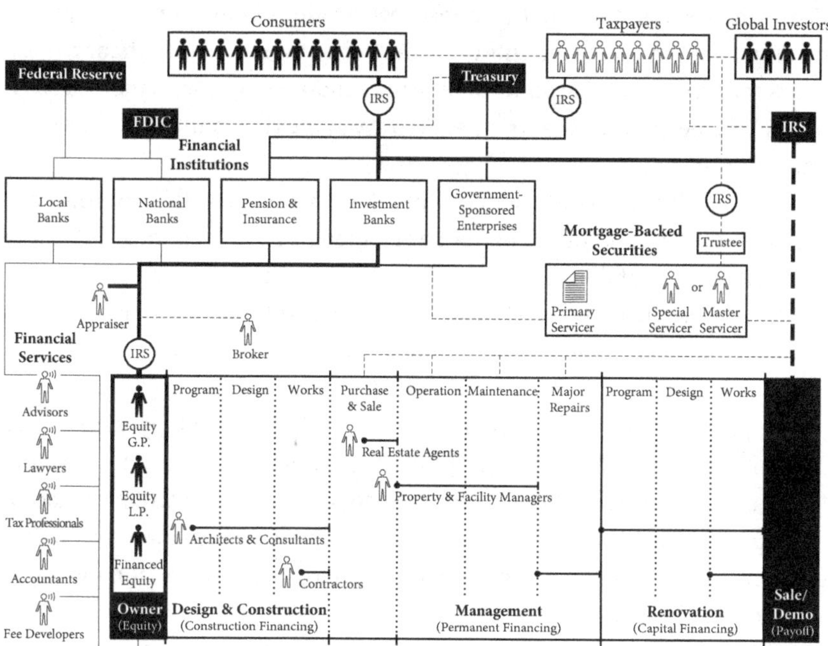

Figure 4.1 The Architecture of the U.S. Real Estate Industry. Design Credit: Oliver Oglesby, Nicholas Chelko, and Jesse M. Keenan.

consumers and investors may see property and asset values decrease, while others may experience substantial increases as their assets become more desirable. Chapters 5 and 6 explore the wide-ranging implications of these emerging market transformations for both sending and receiving zones, respectively. As with most adaptations, there are winners and losers along the way. In the near term, investors are already seeking to capitalize on this gradual movement of people and capital.

This transformation in market behavior is not grounded in speculation. There are already emerging signals across various markets that a shift in behavior is well underway. Property markets for everything from housing to timberland have long been observed to internalize or capitalize risk, such as the risk from flooding[2] and wildfires.[3] Buyers and sellers learn about the underlying risk of various hazards to properties over time, and they reduce the value for high-risk properties. They may also adjust that property value upward for properties protected by risk mitigation infrastructure, such as infrastructure associated with flood protection.[4] This form of hazard mitigation infrastructure is more or less treated as an amenity, whereas the hazard itself is treated as a disamenity. In some cases, these amenities and disamenities shape marginal responses with values being adjusted upward and downward modest amounts often between 1% and 5%.[5] In theory, these figures reflect the fact that some people may choose to live with the risk (e.g., with insurance), while others may seek to avoid the risks entirely by investing somewhere else.

Separating the effect of the environmental hazards from the environmental amenities (e.g., coastal views) can be challenging for economists.[6] Simply living with a view of the ocean or mountains is sometimes worth the risk for people.[7] However, a view of a burned area after a wildfire—even many miles away—has been observed to drive down the value of a piece of property.[8] When climate risks are often out of sight, they are also out of mind. For this reason, planners are now turning to virtual reality technology to visualize SLR, and early experiments suggest that this experience has a demonstrable effect on positively shaping perceptions and adaptation preferences among participants.[9] The downside is that simulation experiments have found that "the number of homeowners interested in moving out of the region increases steadily over time as the sea level rises."[10]

In 2018, a group of young renegade economists, led by the University of Colorado's Asaf Bernstein, published the first peer-reviewed evidence of an observed price signal from climate change in a housing market.[11] They had sought to evaluate whether "risk capitalization" (i.e., internalization of climate risk into prices and values) was occurring in SLR zones. Their model controlled for the distance to coast, elevation, bedrooms, and property types (e.g., condo, single family, multi-family), among many other variables. Their conclusion was that buyers and sellers were accepting on average a 7% discount for properties in SLR zones compared with otherwise similarly situated properties not in an SLR zone.[12] The research also found almost double the discount rate for multi-family properties, suggesting that "sophisticated investors" were increasingly becoming savvy about climate risks and were valuing properties lower (i.e., higher discount rate) as a function of their implied forward-looking expectations of increased carrying costs and/or lower future appraised values.[13]

In addition, this research provided evidence that these "sophisticated investors" may "not [be] sensitive to local beliefs regarding the effect of climate change."[14] Yet, for many market participants, belief systems and perceptions about climate change do matter. The authors concluded "that the SLR exposure discount varies at the county level by the degree to which inhabitants are worried about the effects of climate change: with more worried areas impounding a significant discount and unworried areas demanding no concessions for SLR exposure."[15] Subsequent research would later show that belief systems in local markets significantly shape how buyers and sellers internalize climate risks, if at all. Sometimes, the effect is not limited to discounting behavior. For instance, research has found that overall pessimism about forward-looking values is correlated with lower transactional volumes of up to 20% in high-risk SLR zones.[16] Lower transaction volumes in recent years fly in the face of a very hot American housing market, and it likely signals the cooling off of the market in high-risk areas, such as in Florida.

The effects of belief systems are not just limited to consumers, as some research has found that "[SLR] vulnerability is associated with significantly reduced construction in areas with high climate change beliefs, but . . . this relationship is significantly attenuated in more skeptical areas."[17] Unfortunately, in places like North Carolina that have a mix of climate beliefs, a policy signal that SLR would be incorporated into future land use planning decisions had the perverse effect of driving more housing construction in high-risk areas, as the market was in a race to build before the regulations could take effect.[18] As climate believer jurisdictions in the Northeast consider relocation from the coast, climate denier jurisdictions in the Southeast are doubling down on risky construction.

Although beliefs about climate change are often heterogeneous in the United States, housing and suburban development patterns tend to sort households into fairly homogenous areas.[19] As some economists put it, "Agents derive utility from ownership in a neighborhood of similar agents."[20] A resulting spatial concentration of belief systems was observed to shape behavior insofar as "houses projected to be underwater in believer neighborhoods sell at a discount compared to houses in denier neighborhoods."[21] In other words, communities with lots of climate deniers are not seeing properties decline or slow in value relative to communities who have come to terms with the prospects of physical climate risks.

The political economy of this behavior is significant, particularly because early-stage research suggests that "houses exposed to [SLR] are increasingly more likely to be owned by Republicans and less likely to be owned by Democrats."[22] For Republican-dominated Sun Belt states, local belief systems are likely delaying market adaptations that would rationalize local public-sector hazard mitigation and adaptation investments. The absence of these investments is likely to accelerate out-migration and relocation in the long term. More fundamentally, by not internalizing the risks of climate change, these "denier" communities are facing the prospects of a rapid devaluation when belief systems do change because of either bottom-up sustained local experience with climate impacts or top-down signals from capital markets. This cascading set of institutional and market failures is sometimes referenced as the "climate Minsky moment."[23]

Belief systems are not fixed, and neither are people's responses, even if they do "believe" in climate change. Some research found that even when people are aware of climate risk, they may not necessarily be motivated to take steps to adapt their homes to mitigate that risk.[24] Nonetheless, beliefs about climate can and do change.[25] Targeted advertising about climate change designed to reach Republican audiences has been shown to be partially successful.[26] Simulation experiments that allow homeowners to visualize the impacts of SLR found that political conservatives are not "any less inclined to support adaptation" than other people.[27]

For homeowners, research has suggested that, as buyers and sellers become more aware of SLR projections, there is an increase in their discounting behaviors.[28] This underscores the value of state legislation that requires transactional disclosure of prior flood events or the location of high-risk areas, as well as the value of having greater consumer access to climate risk information. In California, homes required to disclose their location in a high-risk wildfire area were observed to sell at a discount compared to those similarly situated homes without the requirement.[29] Unfortunately, even with widespread publicly available information, such as flood maps, "the information in these maps is not fully capitalized in property values."[30]

Belief systems are not the only determinant of discounting and adaptation behavior that internalizes risk. Affluence and wealth also play a role in how people sort out where they are going to live and how they value their purchase prices and rents. Research in Florida found that low-income and minority households are increasingly sorting into high-risk flood zones, where these households have a lower marginal willingness to pay for flood protection.[31] As the researchers noted, "A willingness to pay to avoid flood risk increases with income or wealth."[32] In the trade-off between the near-term benefits of lower housing costs and long-term expected losses, low-income households are inclined to reduce their upfront costs.

In Florida, the near-term benefits of lower housing costs are likely, in part, going to offset increased water and electric utility cost burdens.[33] Southern utilities, such as Florida Power & Light (FP&L), that have resisted the transition to renewable energy disproportionately pass on the risk of natural gas price volatility, which amplifies observed cost burdens.[34] These utilities also have little interest in helping low-income customers to reduce their power bills by increasing their energy efficiency (EE). Across the country, electric utilities spend an average of $2.58 per customer on EE measures. FP&L spends $0.02 per customer.[35] For many owners and renters in Florida, running the air conditioner is a lifesaving and expensive adaptation to extreme heat. Unfortunately, recent research has found that, across the United States, "extreme heat days result in a statistically significant increase in [eviction] filings."[36]

On the other end of the wealth spectrum, purchasers with greater measures of wealth and education attainment tend to discount future climate risks and values, particularly for investment properties.[37] To the contrary, affluent purchasers of second homes in Florida were found to extend a much lower level of risk discounting.[38] This makes sense. Investors and primary homeowners are concerned about long-term value, and the ultra-wealthy have money to burn in acquiring a second home. While these affluent second homeowners might be able to absorb the risk of loss, their collective behavior clouds the market's price adjustment to internalize SLR and flood risk. As researchers have noted, "This means that the proverbial clock is ticking for these communities to proceed with adaptation measures" before "[it] become[s] more difficult to finance [adaptation investments in the event that] market values of coastal properties and the tax base decline."[39]

The growth patterns of America's coastlines suggests that settlements are becoming wealthier and denser with more structures in all regions of the country.[40] The question is how much longer can the wealthiest segment of the population be willfully blind to the private and public costs of their own indifference? Wealthy people might not be as blind as one might think, as many coastal adaptation investments in infrastructure are strongly correlated with home values.[41] For waterfront property owners, these adaptation investments have been observed to pay off by significantly increasing their property values.[42]

In the absence of widespread agreement on at least the general identification of the risks of climate impacts, the informational asymmetries in the marketplace are anticipated to create a variety of distortions. For instance, climate-aware owners may underappreciate their current risk profile because they anticipate selling to climate-unaware buyers in the future.[43] This game of musical chairs raises tough questions about who will be the last person standing. Luckily, more people are beginning to realize the broader implications of climate impacts on housing. A 2021 survey conducted by Redfin found that one in five Americans believes that climate change is already negatively impacting their property values.[44] Nearly a quarter of respondents in the Sun Belt are already attributing climate impacts to declining values.[45]

As popular attention begins to turn to climate change, so too does the capacity of the housing market to price climate change. One research study found robust evidence of this increasing popular awareness by deploying a "systematic textual analysis of the for-sale listings to measure the frequency with which climate-related text (e.g., mentions of hurricanes or flood zones) appears in the written description of the listed properties."[46] For instance, for those living on high ground out of a flood or SLR zone in Miami, the research found evidence that they might be inclined to mention that in their listing.

From coast to coast, the local evidence is becoming more robust. In Miami, home price appreciation in SLR zones is significantly underperforming properties at higher elevations, despite the influx of wealthy buyers.[47]

In Hawaii, properties in SLR zones "have already experienced declines in transaction prices, at 9 to 14%, attributed to expectations of exposure to chronic inundation."[48] In Hilton Head, South Carolina, properties in high-risk zones have seen a 15% decline in value.[49] In many studies, it is not just properties that are in high-risk zones (e.g., SLR zones) that are bearing the burden of market adaptations. Often, nearby properties that are dependent on vulnerable infrastructure but are otherwise out of a high-risk zone are also being discounted.

Despite this localized evidence, other research has found limited overall price effects of SLR in some of the places most at risk,[50] which raises questions about the role of public policy and even the media in shaping local beliefs.[51] Some have argued that subsidies for everything from high-income tax benefits to beach nourishment have clouded a clearer perspective for consumers, and, without these subsidies, the decline in coastal property markets would be even greater than is currently observed.[52] Despite the noise, somehow the word is getting out. Other properties near the coast—but outside of an SLR and flood zone—are already seeing price appreciation premiums.[53] People in the know are already moving to higher ground.

Depending on the economists' preferred discount rates, houses in flood zones are estimated to be overvalued somewhere between $44 billion[54] and $187 billion.[55] Some coastal areas are conservatively estimated to be overvalued by up to 13% over current housing values.[56] More experimental research places the upper boundary of unpriced climate-amplified flood risk at more than a quarter-trillion dollars.[57] With a housing market worth approximately $47.5 trillion,[58] an approximate .5% overvaluation might not sound like a big deal, but much of the overpriced property is heavily "concentrated in counties along the coast with no flood risk disclosure laws and where there is less concern about climate change."[59]

Aside from the vast implications for housing, there is also increasing evidence that commercial real estate is internalizing climate risk, with "commercial owners/investors in some geographies . . . placing a higher risk premium on all properties in metro areas affected by climate events, regardless of whether their individual properties have been directly affected."[60] Research has found that "after controlling for property size, age, location, time (market conditions), and occupancy, . . . hurricanes appear to have a significant impact on property values, appreciation (net of capex), and total return."[61] In major markets such as NYC and Boston, commercial properties in high-risk flood and inundation zones have seen persistent increases in capitalization rates (i.e., lower comparative asset values), even among properties that have not previously been flooded.[62] There is also early-stage evidence of the negative relationship between commercial real estate returns and emerging heat stress.[63]

For U.S. equity real estate investment trusts (REITs), climate impacts have been observed to heighten return volatility.[64] Other research has found that REITs with high measures of physical climate risk exposure "tend to have lower cash flow and firm values."[65] These changes in valuation of commercial properties are being facilitated by changing professional standards for property appraisals that are actively incorporating physical climate risks into their methodologies.[66] In this sense, discounting climate risk is not just about beliefs—it also requires professional skills. However, even in commercial real estate, belief systems do matter, with early-stage research "suggest[ing] that higher disaster risk is associated with higher cap rates [i.e., lower values], with the effect being significantly more pronounced in areas with a stronger belief in climate change."[67] When commercial values decline and mortgage credit tightens, the implications for small businesses, local tenants, and even a local tax base can be significant. Any exit of capital will make high-risk areas more expensive to maintain and will limit their ability to make adaptation investments.

Mortgage Markets

When it comes to commercial mortgage lending, it appears that originators are increasingly of the belief that climate change represents a material risk. They are careful about what mortgages they keep on the books and what risks they retain when they sell mortgages to the capital market. A recent study found that when mortgage originators retained some of the default risk for the mortgages they sold to the capital markets, they "carefully assess[ed] climate risk and avoid[ed] exposure to climate hazards," and the "default risk associated with climate hazards significantly decrease[d accordingly]."[68] Default risks are a real problem for commercial lenders, with research showing, for example, that "both [Hurricanes] Harvey and Sandy led to elevated levels of commercial mortgage delinquency."[69] The good news is that lenders appeared to adapt to these emerging risks by incorporating flood information into their underwriting processes.[70]

While the commercial mortgage market in the United States is around $5 trillion,[71] the residential mortgage market is more than $14 trillion and represents a much broader structural challenge for underwriting climate risk.[72] All mortgage investors face one or more physical climate risks.[73] The first risk is that the collateral real estate goes down in value, at which point the property might be worth less than the outstanding amount due on the mortgage loan. Mortgage lenders typically like to have a loan-to-value (LTV) ratio of somewhere between 60% for commercial and 80%–90% for residential loans. As the LTV narrows with declining values, then the risk for mortgage investors goes up. Because most residential mortgages have fixed interest rates, there is little that a lender can do after the fact.

Another risk that mortgage investors have is default risk, wherein the borrower simply stops paying.[74] Sometimes this is just a temporary cessation in payments where the borrowers become delinquent while they try to repair their properties. In other cases, this happens when properties are destroyed and borrowers lack the resources or incentives to keep paying the mortgage. The final major risk that mortgage holders face is prepayment risk.[75] This most often happens when a property is destroyed, and the borrower uses the insurance proceeds to pay off the balance of the mortgage loan. In this case, mortgage investors do not get the full benefit of interest payments that they would have otherwise received without the destruction of the property.

Mortgage lenders have two fundamental ways that they can manage and adapt to these climate risks. First, lenders can charge higher interest rates (i.e., "climate risk premium") and offer loans for shorter terms, with lower LTVs under conditions of tighter credit and full insurance coverage. They can also outright reject mortgage applicants and pursue lending outside of high-risk areas. Second, they can pass along these climate risks by selling mortgage loans to investors in the capital markets. This process is known as *securitization.*

Researchers have found residential mortgage lenders are charging higher interest rates on properties in SLR zones.[76] But this climate risk premium is "smaller when the consequences of climate change are less salient and in areas with more climate change deniers."[77] Other research found that mortgage lenders are tightening lending standards for prime mortgages in post-disaster counties where there is a strong belief in the negative impacts of climate change.[78] Unfortunately, "lenders do not [appear to] update their risk assessment in low [climate change] belief counties."[79] Sometimes it simply boils down to loan officers themselves. Recent findings found that "loan officers approve fewer mortgage applications and originate lower amounts of loans in abnormally warm weather. This effect is stronger among counties heavily exposed to the risk of [SLR] . . . and for loans originated by small lenders."[80] It is not just today's warmer weather; research has also found that "interest rates are higher and loan terms are shorter in areas forecast[ed] to experience a larger increase in the number of hot days over the coming decades."[81]

Beyond the modification of mortgage loan terms, lenders can also pass on climate risks by selling mortgages to third-party investors or even the government-sponsored enterprises (GSEs) known as Fannie Mae and Freddie Mac who act as a conduit for the capital market. While the GSEs are technically independent of the federal government, they have been under conservatorship of HUD's FHFA since the Great Recession, where they are likely to remain for the foreseeable future. Despite years of turmoil, the GSEs still retain roughly half of the mortgage market in any given year. As was previously referenced in a commercial context, residential mortgage originators may be obligated to retain some of the risk in the mortgages they pass on to the capital markets, particularly through the GSEs. This risk retention innovation brought about by the Dodd-Frank financial reforms enacted after the Great Recession was intended to impose discipline on originators hungry to profit from high transaction volumes.[82] For all intents and purposes, banks profited from the upfront fees and passed on the risks to the federal government. As the originators got sloppy with processes like verifying income, the risks piled up for the GSEs. This is a classic example of a moral hazard.

Unfortunately, the GSEs do not appear to be assessing physical climate risks that might otherwise mitigate another moral hazard. In the context of coastal and hurricane risks, researchers found that the GSEs are treating inland and coastal area risks the same despite the large differences in risk exposure.[83] More alarmingly, their models suggest that if the GSEs were to allow for the inclusion of catastrophic risk in assessing overall credit risk, then private investors would significantly reprice their investments in high-risk coastal zones. The authors noted that "by preventing markets from pricing mortgage credit risk heterogeneously across locations, the GSEs prevent the internalization of climate risks.[84] When private investors reprice mortgages, the problem is not just that mortgages get more expensive. A paper by the economists Matthew Kahn and Amine Ouazad found that "as risk increases lenders pivot towards higher FICO [credit scores] and higher income households."[85] Their answer to the inevitable problem of credit constraints in high-risk areas is to double down on the design of GSE securitization products. By spatially diversifying risks and incentivizing local adaptation investments among borrowers, their argument is that coarser measures of market adaptation that would have otherwise removed credit for both at-risk and lower-risk properties might be avoided.[86] For their idea to work, the GSEs will need to come to terms with climate risk, and that is unlikely to happen during the second Trump administration—even if they are fully reprivatized.

Keeping the money flowing is consistent with recent research that suggests that even among pessimistic agents (e.g., borrowers and lenders) who believe climate is going to impact the collateral sooner than later, the adaptation preference of borrowers is to extend the maturity of the loans in order to reduce the payments and use the lower payments and higher leverage to offset the costs of insurance coverage.[87] Unfortunately, this model of adaptive behavior may be limited by the proposition that insurance might not be available in the future for such high-risk properties. What may be adaptive for mortgage originators might be maladaptive for mortgage investors who get stuck with bad loans. In this light, the limits of securitization to broadly spread climate risk remain largely untested.

Even well-intended securitization designs can have unintended consequences. Fannie Mae's $95 billion green bond program—once the largest green bond program in the world—was designed to support investments in sustainable housing construction. But the program received extremely poor ratings for a lack of physical climate risk assessment from its third-party green bond certification provider, CICERO.[88] Investing in sustainability and resilience does not make a lot of sense if the housing is located in areas that will soon be underwater or burned to the ground.

To their credit, the GSEs attempted to correct this oversight by investing in the staff necessary to support climate risk management activities. However, without clear direction from the FHFA and Congress, they ultimately determined that explicitly identifying high-risk properties or even charging a climate risk premium would be politically and legally challenging. As will be discussed, it is also technically challenging. For now, the GSEs are leaving it up to investors and the credit rating agencies to figure out. Unfortunately, figuring out the true nature of the risk is going to be even more difficult. After indirect pressure from the second Trump administration, the GSEs disbanded their in-house climate units, and that leaves homeowners and investors further in the dark.

While some banks are largely operating in the dark about climate risks, others are not. Interviews conducted over the past decade with banks and mortgage lenders have found that some banks have made a strategic decision to avoid mortgage lending in certain high-risk areas. If they do lend, they are sure to securitize these mortgages with the GSEs and/or the capital markets. This adaptation in the financial services industry is known as *bluelining*.[89] The term originated by the observation of a banker who drew a line around a particular flood zone shaded in blue.

One has to acknowledge that there are two spatial realities for pessimistic lenders who want to manage climate risk. There is an area where they have the analytical capacity to estimate and project climate risk. In this area, they have the capacity to take actions to manage their risk (i.e., Actionable Risk Zone). However, another area exists that represents the true area of risk from climate impacts (i.e., True Risk Zones). As represented in Figure 4.2, they are not necessarily overlapping areas.

There are two reasons that explain the potential discontinuity of Actionable Risk Zones and True Risk Zones. First, the mapping and climate assessment technologies are not as refined as one might think. Figuring out precise estimates of various climate impacts and hazards on a property-by-property basis is still several years out. Although discrete SLR zones are helpful, other impacts are not so easily mapped with useful measures of confidence. Likewise, mapping techniques themselves do not easily graft onto underwriting and portfolio-management systems. As such, lenders often use coarse approaches to physical risk assessment, such as simply excluding entire census tracts, that lead to a misalignment between Actionable and True Risk Zones. This discontinuity between the zones may lead to distortions and unfair outcomes wherein borrowers who are not in a True Risk Zone are penalized for being in an Actionable Risk Zones, and vice versa. In theory, advances in science and technology will drive a convergence between these two zones over time.

The second reason for divergence relates to the strategic game playing of the lenders themselves. With imperfect models, a lack of local information, and an absence of regulatory oversight, lenders are performing a balancing act between managing risk on one hand and still maintaining market share and transactional revenue on the other hand. However, as climate risk assessment models become more precise, reliable, and standardized, this is likely to provide the objective data points necessary for regulatory oversight, particularly among banking regulators.[90] Banking regulators are increasingly concerned that a concentration of mortgage assets in regional

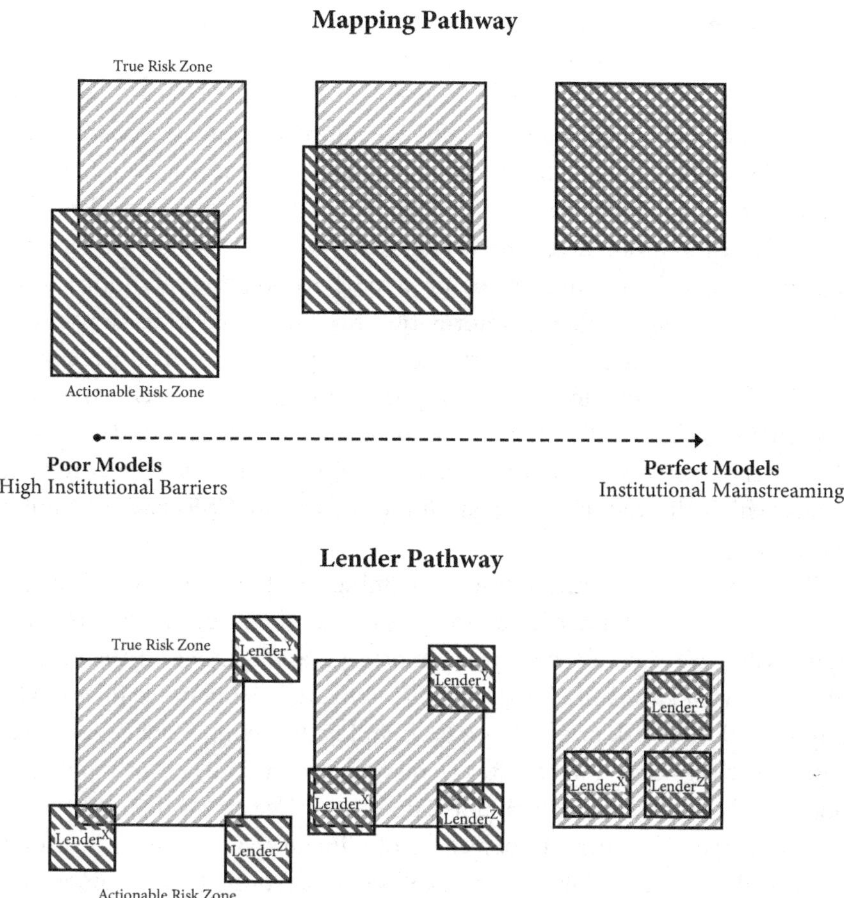

Figure 4.2 Pathways for the Bluelining of Mortgages and Financial Services. Design Credit: Oliver Oglesby, Alper Turan, and Jesse M. Keenan. Data Source: Keenan, Jesse M., and Jacob T. Bradt. 2020. "Underwaterwriting: From theory to empiricism in regional mortgage markets in the US." *Climatic Change* 162(4): 2043–2067. https://doi.org/10.1007/s10584-020-02734-1

banking systems (e.g., Florida) may represent a material sub-systemic risk to financial stability.[91]

Not all banks are flying blind. Novel bluelining research tested a simple premise—local banks know what is coming.[92] Banks come in all sizes, from large national banks like Bank of America to local banks with only one or two branches. Researchers speculated that small banks would have better soft information about emerging climate risks. While national banks operate with complex systems for mortgage origination and underwriting, local banks often underwrite loans with the benefit of knowing the community. Loan officers do not need to be told that a property is at risk from flooding if they drive by that flooded property on their way to work every day. This is in contrast to online fintech mortgage lenders who automate underwriting and who are increasingly recognized to be overlooking soft information about climate risks. This soft information has been used to the advantage of local banks in the past. For instance, leading up to the Great Recession, small banks were able to exit the subprime mortgage mess well before the large national banks did. This initial bluelining research found evidence that smaller banks were much more likely to securitize mortgages in SLR zones, which allowed them to pass off the risks to the GSEs and the capital markets.[93]

The broader implications from bluelining are that lenders may begin to tighten credit standards, more critically evaluate collateral property appraisals, and possibly even deny mortgage applications in the face of heightened measures of climate risk assessment.[94] Sometimes they might impose a climate risk premium on households that are in True Risk Zones, and sometimes they might entirely miss the mark and unfairly penalize households who are not in immediate risk, at least within the term of the mortgage. The current coarseness of climate risk assessment raises a great deal of concern for the maladaptive implications for consumers. As a consumer advocacy group noted:

> Early signs of bluelining appear to overlap with the same patterns of disinvestment established by redlining. As financial institutions increasingly factor climate risk mitigation into their business strategies, low-income communities and communities of color may face a resurgence of racial and economic financial exclusion that mimics redlining practices, with increasingly debilitating consequences as the climate crisis intensifies.[95]

Bluelining has been conceptualized to have a strong spatial correlation with historic patterns of redlining to the extent that historical patterns of housing disinvestment also resulted in infrastructure disinvestment by virtue of a degraded property tax base.[96] This condition exacerbates the current challenge of managing climate impacts (e.g., extreme precipitation) with low quality infrastructure that often can barely keep up. Just ask the redlined communities of New Orleans whether their stormwater pumps are able to keep up with run of the mill rain events—they can not.

Research has consistently highlighted the overlapping spatial boundaries of redlined, flood,[97] and SLR zones.[98] This spatial convergence, together with emerging evidence that low-income and minority households are sorting into high-risk flood zones, sets up a wicked concentration of socioeconomic vulnerabilities and physical exposure.[99] If bluelining drives another wave of disinvestment in these communities, they may not have the financial capacity to make the critical adaptation investments necessary to keep their communities afloat. This may be one of the single greatest push factors driving climigration and relocation in the United States.

By adapting their financial services activities through bluelining, financial institutions may also run into legal hurdles. In the United States, banks are prohibited from using attributes such as race, gender, and marital status, in evaluating the provision of credit and mortgages products pursuant to the Equal Credit Opportunity Act (ECOA).[100] It is not necessary to show direct discriminatory intent in enforcing ECOA.[101] Rather regulatory agencies may find ECOA liability through the demonstration of "disparate impact [that] occurs when a creditor employs facially neutral policies or practices that have an adverse effect or impact on a member of a protected class unless it meets a legitimate business need that cannot reasonably be achieved by means that are less disparate in their impact."[102]

The legal question is whether a lender's market exit—as a means to manage climate risk—is reasonable in light of the disparate impacts it may cause? Is there an alternative way to manage the risk? It might be a legal stretch to argue that banks should make adaptation or hazard mitigation investments in high-risk zones instead of retreating. These activities may not be reasonable because they might not be profitable in low-income communities. It is in this context that banking regulators recently extended Community Reinvestment Act (CRA) credit to banks to incentivize—but not require—banks to make resilience investments in low- to moderate-income communities.[103]

Bluelining and the disparate impact standard under ECOA raise a significant potential public policy challenge. In this case, the adaptation of banks to manage mortgage lending based on objective climate risk criteria may result in disparate impacts that look a lot like historic patterns of discrimination, given the spatial convergence of low-income and minority populations in high-risk flood and SLR areas. The flip side of the equation that few community advocates and politicians care to acknowledge is that restricting access to mortgages in some high-risk areas may result in long-term increases in social welfare through the avoidance of devastating future losses. In the near term, a lack of transparency about market behaviors may not send adequate signals to potential homeowners and renters in a manner that informs their risk-adjusted behaviors—further exacerbating measures of economic and racial inequality.

It is not quite a "take your medicine now or later" problem, as many people may not appreciate the chronic conditions that they face. Climigration and relocation may be the only options that some people have in the midst of bluelining and the exit of capital in high-risk communities. Unfortunately, the lack of regulatory oversight is allowing an uneven landscape where mortgage capital flows into some high-risk areas and away from others. As the financial services sector seeks to adapt, they are running up against wicked problems that have accumulated over generations of poor public policy decisions. With the public sector lacking the capacity to properly plan for adaptation in the built environment and to supervise the adaptation of financial services, the private sector is increasingly defining the landscape for the future of post-climate housing.

Municipal Bonds and Public Finance

As a general but not universal proposition, when one sees a shift in the mean observations of climate change (e.g., temperatures or precipitation), there is also an observed flattening of the statistical distribution insofar as the variance in the distribution reflects a greater range for potential extreme events.[104] In the parlance of finance and engineering, this phenomenon is associated with a greater likelihood of "fat tail" extreme events. For example, the ARI for storm surge on the scale of Hurricane Sandy in NYC that caused billions in infrastructure damage was approximately a 1-in-1,200 year event in the pre-industrial period, and has since been estimated, in the face of climate change, to be somewhere between a 1-in-23 and 1-in-93 year event by 2100.[105]

A greater likelihood of extreme climate events means a greater measure of financial risk for infrastructure investors and owners, including public and private utilities and local governments. Market participants and engineers both utilize a statistical understanding of climate impacts in managing financial risk and designing infrastructure systems.[106] Municipal (muni) bond investors have to evaluate a range of contingencies and risks in underwriting the performance of an investment that could shape their financial returns over a relevant hold period. Climate analytics firms in the financial services sector play an increasingly important role in evaluating a potential risk and return frontier for any investment portfolio.

As a general proposition, when there are greater risks, there must be a corresponding increase in financial returns. In the context of growing physical risks from climate change, this plays out in the notion of the previously referenced "climate premium" in debt and equity markets. Investors demand a higher rate of return to offset a particular risk. Although this research is at an early stage of development, climate premiums have been observed not only in real estate and mortgage markets,[107] but also in sovereign bond markets,[108] derivative markets,[109] and corporate bond and equity markets.[110] A variety of factors, including uneven corporate disclosures, have clouded consistent empirical research in understanding emergent investor behavior in pricing climate risk in debt and equity markets.[111] However, research suggests that markets are getting more efficient at pricing climate risks across a variety of asset classes as their climate intelligence capacity increases.

One outcome of these market adaptations is that the muni bond market is beginning to price climate risk. Recent research found that jurisdictions with greater levels of physical risk are paying higher prices for the cost of their muni bonds.[112] More precisely, this research observed a 23-basis-point increase in annualized cost for every 1% increase in measured physical risk. In the context of an average bond issuance of $27.5 million, this climate risk premium starts at an annual cost of $64,000 and goes up as the risk increases. A similar effect of a 15-basis-point annual cost premium was projected for local governments subject to economic damages from extreme heat stress.[113] Similar risk premiums in the muni bond market have been found for flood, drought, and wildfire risks.[114] Although these figures may not sound cost prohibitive, it is widely estimated by market participants that such premiums will increase over time as municipal bond disclosures increasingly incorporate physical risk considerations at the behest of credit rating agencies and institutional investors.[115]

Other research has found that the muni bond market has been pricing SLR exposure since at least 2013, with the observed price effects stronger for longer-term bond maturities.[116] The longer that investors have to wait to get paid, the greater the risk as sea levels rise. It is not just longer-maturity bonds that see an amplified effect, as "revenue-only and lower-rated bonds" also trigger greater scrutiny for compounding risks.[117] In an interview a decade ago, a Wall Street bond buyer noted how he unceremoniously passed on buying a Key West Aqueduct Authority bond issuance after looking at their facilities in NOAA's online SLR viewer platform.

The emerging risk premiums have at least two major implications for utilities and local governments. First, higher-risk jurisdictions should anticipate a higher weighted average cost of capital (WACC) for projects, and, second, jurisdictions that take steps to mitigate and adapt to climate change may be rewarded for their efforts with lower rates, as has been recently observed.[118] Infrastructure finance is often a piecemeal process that utilizes a variety of market and below-market credit products, as well state transfers and grants. Every dollar counts. There is little room in the current domestic infrastructure finance system for climate premiums.

This raises a fundamental question: What are investors and credit rating agencies really worried about? When local governments and infrastructure providers issue general obligation bonds, investors are concerned that a degraded tax base may impact the fiscal stability of the issuer over the term of bond.[119] Research led by Cornell urban planning scholar Linda Shi found that property tax revenue in the rocky shoreline state of Massachusetts could be impacted by SLR by up to 25% for a few select local governments,[120] while a review of more than 400 local governments in the lower-elevation state of Florida were projected to suffer an average loss of local revenue of just under 30% with SLR.[121] This is just a single measure of direct exposure to SLR inundation and does not account for future accelerated devaluation of local properties associated with discounting by and between buyers and sellers.

Recent research going back to 1980 of more than 2,000 local governments on the Atlantic Coastline, found that hurricane strikes increase the risk of muni bond defaults over the subsequent decade.[122] More recently, Paradise, California, has been teetering on default of its bond obligations after a wildfire destroyed most of the town and its infrastructure.[123] Even Los Angeles faced a negative credit outlook after the wildfires in 2025. In 2024, Clyde, Texas, did actually default on its bond, and it was evaluated by Municipal Market Analytics as the 12th such default partially attributed to climate impacts since 2009.

Default risk is amplified for revenue bonds, which are bonds that are paid off exclusively by revenue from particular projects.[124] For instance, a toll bridge to a low-elevation barrier island financed by revenue bonds might be a risky investment if fewer cars drive over the bridge as the island is gradually inundated by SLR. To compound the revenue bond risk, recent research has highlighted that after disasters, local governments are more likely to issue revenue bonds and that these post-disaster issuances are paying higher comparative yields.[125] Climate-fueled disasters are driving up risks and costs.

It is not just climate impacts that are driving up costs; state policies themselves can drive up costs. Municipal bonds are rated by rating agencies, who evaluate everything from the credit quality of the issuer to the climate risks facing capital assets. These ratings are usually necessary for the issuance of a bond, as potential investors need to evaluate the underlying risks to determine the price of the bond. In 2023, lawmakers in Florida proposed legislation that would prohibit local governments from using rating agencies that utilize environmental, social, and governance (ESG) factors in determining their ratings. For Pasco County, Florida, the mere threat of such legislation caused sufficient uncertainty in the status of their pending ratings from Moody's and Standard and Poor's that it forced them to delay their issuance of a $345 million capital improvement bond. During this delay, the borrowing costs went up a quarter point, which ended up costing the county an estimated $12,750,000.[126] Even political rhetoric has a cost.

Additional concerns for both owners of and investors in infrastructure relate to the increased direct costs of climate impacts. As previously referenced, the borrowing costs in the muni bond market have been observed to be increasing in response to physical risk assessment, and this may increase an overall project WACC. Increases in capital costs may also be attributed to increased construction costs for incorporating engineering resilience elements within particular assets.[127] For instance, a flood protection system may have to build a higher levee with greater pumping capacity to accommodate future projected precipitation, storm surge, and SLR performance standards.[128] Unfortunately, there is almost zero cost estimate research for these incremental costs across infrastructure asset classes that could otherwise guide local governments in preparing long-term capital plans.

Beyond incremental capital and construction costs, operations and maintenance (O&M) costs from climate impacts are perhaps one of the most immediate cost burdens for infrastructure providers today. Even without climate impacts, infrastructure is very expensive to maintain, with any given system requiring annual O&M costs anywhere between 1% and 10% of the original construction cost. Additional costs for things like repeated road resurfacing in areas that face more ice than snow with warming winters[129] or greater energy costs for pumping more stormwater[130] are all drags on the financial balance sheets of infrastructure providers, particularly in the context where many utility ratepayers are already cost burdened.[131]

Unfortunately, infrastructure managers have a limited amount of convergent research and practice to support the utilization of lifecycle cost accounting (LCCA) models to estimate the total costs over the lifetime of an infrastructure asset impaired by the shocks and stresses of climate impacts.[132] LCCA models are useful for adding up all the present and future costs of building and maintaining an asset. It helps owners plan for maintenance and renovation costs. If extreme heat is going to increase energy costs in the future, owners will want to know this in order to figure out the overall financial feasibility of an asset. Compounding this problem, the United States is grossly behind in the codification of climate-sensitive design and engineering standards that ensure reliability, security, and the engineering resilience performance of infrastructure systems.[133]

Collectively, these additional unanticipated O&M costs—for everything from higher energy costs to post-disaster recovery costs—may steer money away from capital improvement plans. They may also exacerbate ongoing deferred maintenance challenges that not only threaten service reliability but may also reduce the overall useful life of an asset.[134] As the costs go up, this also means less money that could otherwise go to adaptation investments. Some infrastructure owners are already planning for the strategic decommissioning of infrastructure to save long-term capital and O&M costs. In places like Miami-Dade County, Florida, that includes strategic efforts within capital plans to concentrate infrastructure and development on high ground.[135]

Infrastructure Management

The failure or degradation of infrastructural systems provides an array of transmission mechanisms for a variety of economic and financial costs. These are not just abstract long-term projections. These are costs paid by taxpayers, consumers, ratepayers, and a broad range of market actors. While the academy's understanding of a range of impacts and costs across various infrastructural sectors is an active area of inquiry, it is useful to highlight a few critical infrastructure sectors, including the energy, transportation, and water and wastewater sectors.

In the domestic electric power sector, various research has estimated that electricity demand will increase somewhere between 2% and 4% each year in the coming years, with significant demand associated with data centers. However, it has been projected that just heat impacts alone in the United States would increase production costs on the order of $50 billion a year going into the year 2050.[136] As demand increases for electricity, so too will the costs. While the heating and (increasingly) cooling of buildings is a major demand factor, climate change will also increase electricity demand for vehicle charging, wastewater recycling, pumping drinking water farther distances, water desalinization, increased agricultural irrigation, and even water heating, as people tend to take more showers on hot days.[137] Additional unanticipated post-disaster recovery costs from extreme events ranging from hurricanes to wildfires are among a wide range of direct costs that will continue to be passed on to utility ratepayers.[138]

Luckily for ratepayers, electricity retail rate projections by researchers at the DOE suggest that some of these increased costs could be partially offset by decreases in fuel costs stemming from the current renewable energy transition currently well underway in the United States.[139] Unfortunately, these benefits may not be fully captured in Sun Belt states that are beholden to fossil fuel interests among their regulated utilities. Even in a best-case scenario with a post-Trump Democratic White House, the costs will go up before they go down with the termination of IRA tax benefits for renewable energy generation.

In the transportation sector, physical impacts have been widely observed for everything from extreme heat waves to flood events compromising roads, tarmacs, pipelines, and rail lines, with direct repair and delay costs being distributed throughout the economy.[140] Annual direct damage costs for road and rail impacts alone are projected to be just under $20 billion a year by 2050 under an RCP 4.5 scenario,[141] with approximately $1.5 billion in direct costs for bridges.[142] Under the same scenario, increased O&M expenses for the paving and resurfacing of roads are projected to be approximately $19 billion a year by the end of this decade.[143] None of these cost projections account for the indirect costs to consumers and market participants for increased logistical costs, fuel costs, delay costs, and lost economic output.[144] For hourly wage workers stuck in flood-induced traffic congestion, that is money out of their pockets.[145] Current estimates for flood-induced traffic congestion on the East Coast exceed 100 million vehicle hours a year, with that figure going up to 3.4 billion vehicle-hours a year by 2100 under a modest SLR scenario.[146]

The economic impacts for the water and wastewater sectors are projected to be significant. Annualized urban drainage infrastructure damages are projected to be approximately $4.3 billion a year by 2050 under RCP 4.5, with an additional $2 billion in costs associated with water quality and quantity impairments for existing municipal and industrial water supply systems.[147] However, this latter figure does not account for the untold cost of adaptations to addresses water scarcity challenges facing the Southwest, where reduced irrigation, additional groundwater mining, and increased reservoir storage capacity pose significant costs.[148] According to some estimates, the Southwest faces upward of $1.4 trillion in lost GDP by 2050 due to the risks and costs of projected water scarcity challenges.[149]

For taxpayers, the economic consequences of climate impacts are particularly palpable when there is a convergent vulnerability in infrastructure systems, as was highlighted by the landfall of Hurricane Ian in 2022. At the epicenter of the landfall in Lee County, Florida, direct infrastructure repair cost estimates now exceed $320 million, with another $288 million needed for future hazard mitigation.[150] Although these infrastructure costs equal approximately $771 for every resident in the county, it is the federal government that is disproportionately footing the bill for billions in grants and loans to expedite reconstruction of this infrastructure across Florida. It is ultimately taxpayers and infrastructure consumers who will foot the bill for infrastructure recovery, risk mitigation, and climate adaptation. To make matters more challenging, if the second Trump administration is successful in ending federal disaster aid, it could very well stimulate a wave of local bond defaults and municipal bankruptcies that further drives up costs and drives out local residents.

Following Hurricane Ian, research in Florida found an increase in favorable attitudes for climate action and climigration that cut across political, education, and age attributes of local survey participants.[151] While these effects would eventually wane over time, many people did not have the resources to rebuild in high-risk areas, particularly among the high concentration of retirees in the population.[152] As such, many local residents and businesses have already been observed to have moved inland off the coast.[153] Those with the resources are being lured inland to wealthy development enclaves like Babcock Ranches that are packed with their own private infrastructure and resilience amenities that are marketed as places of refuge that will not lose power.[154]

At the center of the state, far from the coasts, Ocala was the fourth-fastest-growing metro area in the country in 2024—growing 3.4% alone in the year after Hurricane Ian.[155] It is now larger than the state capital of Tallahassee. With the lowest estimated level of flood risk of any Florida city,[156] lower comparative measures of climate risk and cheaper insurance are leading motivations for newly relocated residents and businesses. According to Kevin Sheilley, CEO of the Ocala Metro Chamber & Economic Partnership, "We see it as a key competitive advantage."[157] More than 50,000 units of housing and 86,000 home lots have been approved in recent years in the surrounding county of this once rural outpost.[158] When there are local and regional shifts in population (e.g., moving inland) over time, this runs the risk of supply and demand imbalances that could lead to utility rate shocks for receiving zones and sending zones. Luckily for Ocala, their nonprofit public electric utility has active solar net metering and community solar programs. Their electricity bills actually went down approximately 8% in 2024. Sometimes moving north simply means moving north to Central Florida.

Researchers have begun early-stage research estimating domestic climigration and the extent to which the movement of people and industries to receiving zones will create supply and demand imbalances in civil infrastructure systems.[159] As the researchers note, "Climate change is expected to shift this pattern [of population settlement] north and east, contributing to land use change and new demand for civil infrastructure in places where past development strategies might not be regionally appropriate."[160] The combination of fiscal shocks and out-migration of local ratepayers and taxpayers is already a substantive concern for jurisdictions that struggle to stabilize populations following disasters. For those who do not have the option to relocate out of sending zones, the costs are likely to go up, as there are simply fewer people to foot the bill. For instance, the local electric utility Entergy in New Orleans only a few years ago removed a billed surcharge for post–Hurricane Katrina reconstruction—13 years after the fact.[161]

Insurance

Interviews conducted over a decade ago on Miami Beach with low-wage service workers uncovered a curious finding. They were moving off the barrier island to the mainland of Miami-Dade County. It was not the high rents that were driving them off the "beach"; rather, it was the cost and availability of car insurance. As a barrier island, precipitation has nowhere to go except Biscayne Bay or the Atlantic Ocean. On the bay side of the island along Alton Road, the convergence of high "king tides," extreme precipitation, and SLR resulted in a salty mix of sand and water that would frequently inundate a key north-south artery. Not only would this lead to traffic delays, but it was also inundating cars with highly corrosive saltwater on a regular basis. According to local insurance agents, car insurance companies started incurring increased repair costs and claims, and they responded by raising rates and dropping coverage. For low-wage workers, their vehicles were their primary form of transportation to work, and they had no other alternative but to move to the mainland to get insurance coverage they could afford.

More recently, the insurance industry has sustained billions in auto losses, particularly from Hurricane Ian.[162] Progressive Insurance alone had estimated auto losses in Florida of $574 million in 2022.[163] Florida is now the third–most expensive state in the country for full-coverage auto insurance at $3,430 per year—a full 53% above the national average.[164] At 61% above the national average, Louisiana has the most expensive car insurance in the country at $3,606 per year. This expense is equal to about 8% of the average take-home pay for Louisiana households.[165] Unfortunately, car insurance is merely where the climate story begins.

To understand the mechanics of climate change and insurance, one has to understand the architecture of the U.S. insurance industry. As represented in Figure 4.3, the industry is built on a variety of state and local actors

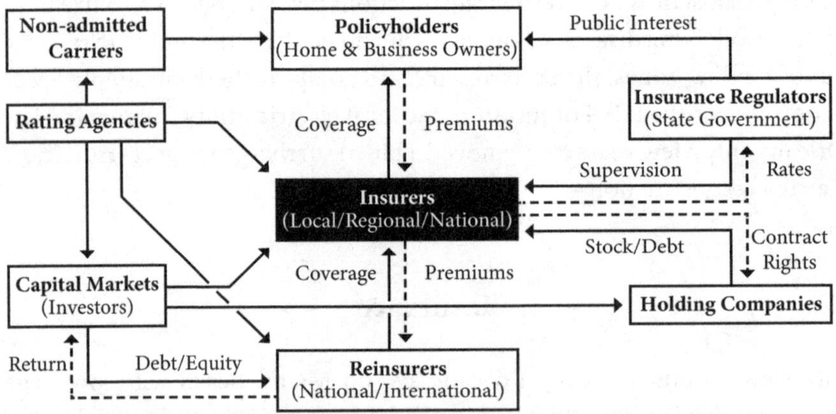

Figure 4.3 The Architecture of the U.S. Insurance Industry. Design Credit: Oliver Oglesby and Jesse M. Keenan.

that extend to a broader realm of international capital market investors. Policyholders buy coverage from one of dozens of large and small insurance companies operating in any given state. Eight states each have more than 100 insurance companies that write property and casualty insurance (P&CI).[166] While large insurance companies often capture the headlines, together they often make up less than 10% of the market share within a state. Many insurance companies are local or regional firms that emerged over time to accommodate local needs, assets, and risks. Almost all of these insurance companies regardless of size will buy coverage for some portion of their risk from reinsurance companies, who in turn are active participants in the capital markets. This transactional cycle happens every year, as most insurance is priced on a year-to-year basis. Historically, this has been understood to be an impediment to pricing the long-term risks of climate change. Unfortunately for consumers, climate risks are now increasingly well represented in the near term.

All states tax and many states regulate insurance companies. Taxes on premiums in 2023 provided $32 billion in revenue for states.[167] Recent proposals have even sought to extend taxes on P&CI premiums to develop a state resilience trust fund that could make investments to reduce climate risks, which would have the reciprocal benefit of stabilizing the exposure of a market.[168] But, even in progressive New York, few policy makers have been much concerned with climate impacts on the insurance market.

The general goal of state regulation is to ensure that the premiums on different regulated lines of insurance coverage are commensurate with the stated, or sometimes regulated, profitability of a firm and to ensure that the firm offering insurance will have the capacity (e.g., credit, cash, reinsurance coverage) to pay claims. In many states, insurance companies must go through an extensive regulatory filing to demonstrate their compliance. In some states like Florida, insurance companies may largely bypass this process if they limit their year-to-year increase in premiums under a certain fixed percentage (e.g., 12%). Any increase in premiums beyond that fixed percentage would require a sometimes lengthy regulatory filing.

In normal years, an insurance company will pay claims from cash reserves and other liquid assets. As a senior executive in one of world's largest insurance companies once noted, "We don't make money on premiums. We make money by investing those premiums."[169] As such, premiums paid into a company are not always in liquid or semi-liquid assets. Regulators often look closely at the capacity of insurance companies to pay claims, because if an insurance company goes under and files for bankruptcy, then those customers will be forced to look to a state guarantee fund for payment. Depending on the state, these guarantee funds are paid for by a mix of insurance companies and consumers through taxes or surcharges. In Louisiana, insurance companies' share of the surcharge is tax deductible. This deduction was meant to attract more insurers to the state. This is how the system works under stable times. As the insurance market faces more stress, consumers sometimes turn to non-admitted carriers, which are largely non-regulated insurance companies that seek to fill the gaps in coverage left by traditional insurers. The biggest downside to these non-admitted carriers is that they are often undercapitalized, and, if these firms go under, then consumers will not have access to a state guarantee fund. As many in Florida are increasingly finding out, they may be left high and dry—or low and wet, depending on how you look at it.[170]

When claims activity is elevated (e.g., active hurricane season), insurance companies may be forced to borrow money to pay claims. Things begin to get tough when these companies have to borrow money over the course of successive years before they have a chance to catch their breath and build their reserves. Investors and shareholders may also put pressure on insurance and reinsurance companies by demanding higher rates of return on their investments to cover the increase in risk. In turn, insurance rating agencies either downgrade or threaten to downgrade the firms under financial stress, which amplifies their overall weakness. Firms with poor ratings may have their activities limited, or they may even be barred from doing business. In a worst-case scenario, a firm may simply become insolvent and leave the state with its baggage. In a recent trend that adds insult to injury, fly-by-night insurance companies sometimes send all their available cash to out-of-state affiliate companies before they go under.[171] Unfortunately, recent research has highlighted that insurance "firms with high [climate] exposure show relatively higher financial weakness."[172]

Elevated claims activity attributed, in part, to climate change are driving a contagion of fear in some parts of the P&CI industry. In 2024, private and public insurers lost more than $112 billion from extreme weather events, which was up from $80 billion in 2023.[173] In 2023, the P&CI industry reported $38 billion in underwriting losses, marking the highest losses in the past decade. As the industry press noted, "Still, the [PC&I] industry managed to post a pretax operating profit of $39.2 billion, with higher investment yields boosting net investment income to $75.8 billion in 2023."[174] The industry managed to make this profit because of a strong economy where they invested premiums. Otherwise, the industry endured a combined loss ratio of 103%.[175] That means more money is going out the door in claims, expenses, and dividends than is coming in from premiums. According to the reinsurer Swiss Re, from 1994 to 2023, the rate of inflation-adjusted losses from natural catastrophes grew by an average of 5.9% per year—more than twice the rate of global GDP growth over the same period.[176] These kinds of upside-down loss ratios are simply not economically sustainable. In the era of Trump tariffs and market turmoil, it raises the prospects of whether insurance companies are positioned to be profitable at all in the coming years.

In the past several years, major insurance companies like Allstate and State Farm have stopped insuring properties in California, and Farmers Insurance and AAA have exited the Florida market.[177] Rapid increases in insurance costs have been enough to not only drive homeowners in Florida to go without insurance; they have also driven some to leave the state altogether.[178] Even the multi-family rental and affordable housing markets are feeling the pain of increased insurance costs, with HUD recently allowing borrowers to double their maximum deductibles.[179]

Unfortunately, no one is really sure how widespread insurance premium inflation really is. There is no national database that tracks insurance for consumers. Without much regulatory authority, the Treasury Department's Federal Insurance Office (FIO) under the Biden administration was largely unsuccessful in launching a voluntary effort to collect coarse data that could have provided consumers with some insight into where and how insurance companies see growing risks, including climate risks.[180]

Led by the Wharton real estate economist Benjamin Keys, researchers figured out a novel way to get access to national insurance premium data without having to corral unwilling insurance companies. They procured 47 million observations of escrow payments from 2014 to 2023. After taking out expenditures for taxes and mortgage insurance, they were able to infer insurance premium payments. Their research found a "sharp 33% increase in average premiums from 2020 to 2023 (13% in real terms) that is highly uneven across geographies."[181] Part of this unevenness is driven by uneven regulation, as well as the market waking up to climate risks that have been unpriced.

Aside from the increased frequency and intensity of extreme events, premium inflation and loss ratios are being amplified by increased levels of economic inflation for everything from home repairs to car parts. In the housing market, material inflation has largely leveled off from the pandemic-era inflation, but labor shortages in construction are continuing to drive inflationary pressure. As the scale of devastation from hurricanes and wildfires increases, it has become increasingly difficult to find, house, and logistically support the labor necessary to rebuild. To make matters more challenging, when Florida enacted aggressive state immigration enforcement policies, it had the effect of significantly limiting what would normally be large numbers of undocumented migrant workers who would come to the state to sort debris and rebuild homes.[182]

State insurance policies designed to protect consumers can even sometimes drive up costs. In California, major insurers leaving the market took issue with the state's prohibition against using climate models to support the establishment of rates for wildfire coverage.[183] Companies could only look at historic loss patterns. This was a surprising policy to many who had followed California's ambitious climate action in the insurance sector. Former Insurance Commissioner Dave Jones was the first insurance commissioner in the country to run climate stress tests on insurance companies.[184] But California's climate policy ambition ran headfirst into powerful consumer advocacy groups. The climate model prohibition was the outcome of pushback from consumer advocates who feared that experimental climate models—that few actually understood—could be used to justify excessive inflation in premiums. Although it is true that few regulators understood these emerging models, there is a regulatory mechanism to externally validate models. For instance, in Florida, proprietary hurricane models used by the insurance companies are interrogated by experts in a nonpublic setting.

Many in the insurance industry had been calling for a national process to validate what they saw as highly experimental climate models that might not be ripe for application. The insurance industry has historically relied on well-developed catastrophe models (CAT models), but the integration of climate models, such as IAMs, represented a leap forward that would be on the cutting edge of science. CAT models and climate models represent two very different mathematical and computational realities. Given the experimental nature of this integration, many argued that it would be premature to make this leap forward given the economic stakes of getting it wrong. By contrast, CAT models have long been externally validated and are considered a major proprietary element for the risk analysis industry.

However, the climate models narrowly focused on wildfire risks that had been developed between scientists and insurers were generally well regarded among a small cohort of experts who intimately knew fire science in California. Indeed, research has found that "firms that rely on coarser measures of wildfire risk charge relatively high prices in high-risk market segments—or choose not to serve these areas at all."[185] By extension, the potential precision of climate-sensitive wildfire models could actually work to reduce prices.

As Harvard's Juliette Kayyem so succinctly wrote in *The Atlantic*, "Insurers are trying to send a message. The government is trying to suppress it."[186] After "seven of the top 12 insurance companies doing business in California . . . either paused or restricted new business in the state," the state insurance commissioner finally relented and allowed the use of climate models.[187] This policy shift came with one important caveat—insurance companies "will only get to do this if they agree to write more policies for homeowners who live in areas with the most risk."[188]

Strong-arming insurance programs to write policies in high-risk areas is precisely what got the U.S. government into trouble with flood insurance. While the NFIP has been historically subsidized by taxpayers in a manner that has led to significant moral hazards for risky development patterns in high-risk areas,[189] there have been major steps taken to bring rates up to actuarially sound premiums that reflect the real unsubsidized risk.[190] Through programs like Risk Rating 2.0 that gradually bring up the rates, some research has suggested that these new "risk-based premiums will yield a positive societal benefit ($10 billion) because they will incentivize household risk-reduction investments."[191] The NFIP is plagued by old maps and poor local models that often do not capture the true extent of flood zones, but there is hope on the horizon with increased investments in measurement science and supporting technology.[192]

As referenced in chapter 2, this optimism is largely tempered by the ultimate conclusion that the designation of flood zones is a political act that is only partially based on the science.[193] The good news is that there is at least one insurance program with the fiscal capacity to pay claims—for now. The bad news for some is that flood insurance premium inflation is driving a reduction in home values. Recent research found that for every $1 increase in flood insurance premiums, there was a corresponding decline in values of between $40 and $250.[194] Of course, many people simply forgo buying flood insurance. Like the climate effects cited in the housing market, people's perception of climate change itself drives their demand for flood insurance.[195]

Like flood insurance, windstorm and wildfire insurance markets are quickly veering toward greater measures of state participation in the market.

Some large states like California, Texas, and Florida have state-run insurance programs that fill a gap where the private market has largely failed to provide coverage. The windstorm market is particularly challenging, with every state along the Gulf Coast having an "insurer of last resort" program. Windstorm coverage protects properties from a wide range of wind events, such as hurricanes, and the coverage is generally purchased separately from one's homeowners insurance policy. While Florida and Louisiana offer cautionary tales on how dysfunctional the wind markets have become, this is not to the exclusion of other coastal states, from Massachusetts to Texas, where homeowners are reporting significant challenges and costs for obtaining windstorm coverage.

In 1992, Hurricane Andrew made a direct landfall with Category 5 winds on the southern coast of Miami-Dade County. As was previously referenced, Andrew led to a mass exodus of residents and to the creation of the state-of-the-art FBC system for regulating the construction of buildings. After sustaining massive losses, the insurance industry began to exit the Florida market. This left lawmakers in Tallahassee extremely unnerved. Florida's economy and tax base were—as they still are today—heavily reliant on housing and real estate. Tallahassee responded with a demand that insurance companies either write wind insurance or they will be barred from writing all other lines of coverage in the state. This was a major gamble. The insurance companies knew this, and they held firm. In the standoff between Tallahassee and the insurance companies, Tallahassee blinked first and created an entity that would later be known as Citizens Property Insurance Corporation (Citizens) as an insurer of last resort. This allowed P&CI underwriters to separate wind coverage from their homeowners' policies and to shed significant amounts of risk onto the state.

By the early 2000s, Citizens had been successful in stabilizing the market. The intervening decade had also seen fewer destructive storms. That changed in the 2004 and 2005 hurricanes seasons with storms like Hurricanes Katrina, Wilma, and Charlie. From 2002 to 2006, the policy count at Citizens would swell from under 250,000 to well over 1.25 million policies.[196] The program would peak in 2012, with almost 1.5 million policies, and significantly trail off over the next decade, with only a few hundred thousand policies during an unusual period of calm without any major hurricanes hitting the state. During this lull, the state's population would grow by approximately 2.7 million people.[197]

As hurricane activity increased in the early 2020s, the policy count would quickly rise to more than 1.25 million. The private market would again pull back from the wind market for a variety of reasons, including extensive fraud among roofing contractors and significant litigation costs. Unique statutory provisions for attorneys' fees and the assignment of benefits led to abusive litigation, with "insurance claims in the state amount[ing] to 6.9% of all claims nationwide, but 76% of the nationwide claims [from] lawsuits."[198] While the Florida legislature would enact litigation reform, it was widely understood that it would take many years for this legislation to have an effect.[199] Outside of the wind market, the PC&I market more broadly was showing signs of stress, with many smaller insurers finding themselves in increasingly difficult financial positions.

Many smaller insurance companies had been borrowing money to pay claims since the start of a more active hurricane period in 2017. To avoid scrutiny from state insurance regulators, they increased their premiums by the statutory maximum without having to go through more extensive regulatory filings. The problem with this strategy was that they should have been raising their premiums by an even higher amount to build reserves and pay off their debt. As a consequence, rating agencies, such as Demotech, began to warn the market in 2022 that some of these companies would face ratings downgrades.[200] A downgrade would be particularly problematic as it would leave many Floridians with insurance that would not qualify for coverage for their GSE-backed mortgages.

For years, legislators in Florida and Louisiana had tinkered with legislation that allowed upstart insurers to utilize lesser-known rating agencies like Demotech, instead of established firms like A. B. Best, Fitch, Moody's, and Standard & Poor's. In Florida, Demotech alone would soon capture more than 50% of the market share for Florida's insurance market.[201] The implication among market observers was that these lesser-known rating agencies might have lower quality standards that would allow weaker firms to participate in the market and hence satisfy conservative Republicans' appeal for more competition. In other words, observers were speculating that ratings "grade inflation" was merely masking structural instability that was coming along with the mirage of free market competition. The biggest problem with grade inflation is that it increases the prospects that financially weaker insurance companies would go under and dump their baggage with the state.

With prices for insurance going up in recent years, Floridians began to take notice. Florida Republicans immediately blamed Citizens for the insurance problems plaguing the state, without taking responsibility for the policies that had fueled the growing crisis. Although Tallahassee had invested in resilience investments for hazard mitigation, they were a drop in the bucket compared to what would be needed to drive strategic adaptation of the built environment in the face of climate change.[202] After all, Tallahassee did pass a bill that stripped references to climate change from state law.[203] Rather, legislators doubled down on subsidizing the market and further clouding market signals of distress. For instance, in full recognition that reinsurer rates were driving a good bit of the premium inflation in the state insurance market, Florida created a $2 billion subsidized reinsurance program backed by Florida taxpayers.[204]

Unfortunately for Florida, unsubstantiated rhetoric would dominate the political landscape. Governor Ron DeSantis frequently proclaimed that Citizens was insolvent. These proclamations even led to a U.S. Senate investigation of Citizens.[205] While the insurer did represent a significant risk to the long-term fiscal stability of the state, it was not insolvent. In fact, in recent decades it had produced significant cash reserves that were often raided by the state. To back up their political rhetoric, Tallahassee would institute "depopulation" policies that would attempt to reduce the growing numbers of policies held by Citizens.

Unfortunately, 95% of the policies that were depopulated from Citizens went to smaller insurers that were covered by Demotech.[206] Later research from economists at Columbia, Harvard, and the Federal Reserve Board of Governors would confirm what observers had long suspected about Demotech. The researchers found, "that Demotech insurers have a much higher likelihood of insolvency" compared to insurers rated by industry leader A. M. Best.[207] In addition, the researchers found that between 2009 and 2022 almost 19% of Demotech's Florida customers had failed and were liquidated, whereas none of the traditionally rated insurance companies were liquidated.[208] To add insult to injury, the cash payments made by Florida to depopulate Citizens were found to mask the true financial position of insurers "operating in the riskiest areas of Florida."[209] State subsidies were making these insurance companies look more profitable than they really were. The implications for this lax supervision of fragile insurers extended to the mortgage market, where the researchers found that "that counties highly exposed to insolvent insurers experience a 4.6 times higher increase in delinquency rate after [a] hurricane than counties less exposed to fragile insurers."[210] Following the 2022 hurricane season, six insurers became insolvent.[211] By 2023, dozens of companies were on a "watch list" for potential financial problems by the Florida Office of Insurance Regulation.[212]

In theory, climate change should impose more rigorous supervision of insurance companies to ensure that they have the financial capacity to endure increased claims as climate impacts accelerate. Greater supervision and less competition would drive up costs and send signals that some areas are simply too risky to insure. As Juliette Kayyem noted, "Insurers are trying to send a message. The government is trying to suppress it."[213] The problem with suppressing the message is that insurers will cross-subsidize losses in one state by increasing rates and lowering claim payouts in other states. This is where the contagion begins.

In Louisiana, the insurance crisis has reached a breaking point for many. Following active hurricane seasons in 2021 and 2022, 12 insurance companies accounting for one-sixth of the market went under, and all 12 companies were rated by Demotech.[214] Following an exodus of insurers after Hurricane Katrina in 2005, the state had long sought to entice companies to enter the market, even if that meant welcoming many fly-by-night, undercapitalized insurance companies. Several of these companies were sending money to affiliate companies while at the same time booking local losses prior to their failure.[215] On paper, they were never profitable. Like Florida, Louisiana has its own Citizens (Louisiana Citizens Property Insurance Company), which has swelled with the failures of small insurers. It also has its own "depopulation" program that continues to offload policies to financially questionable insurance companies.

Louisiana's insurance market failures are imposing significant burdens on the residents of a small and poor state. To compound the cost burden, by law Citizens' insurance products have to be "at least 10% above voluntary market rates."[216] This was intended to disincentivize people from relying on the state's program. But, for many without any other options in the marketplace, this provision is a painful reminder of the state's inability to supervise the market. In 2023, the state elected a former insurance executive who ran on a platform of deregulation as its new insurance commissioner.[217] He won by default after the only other candidate dropped out. It was a job that no one really wanted. In 2024, the insurance commissioner got much of what he sought from the state legislature—making it easier for insurance companies to drop policies and to impose new rates without regulatory approval.[218]

Between economic losses from hurricanes, SLR-exacerbated land loss, and insurance costs, many people are simply leaving the state. Between 2020 and 2024, Louisiana had a net loss of more than 84,000 people.[219] For former Louisiana resident Steve Kissee, his taxes and insurance were higher than his mortgage payments. "It's kind of like having two mortgages in one," he told *The Wall Street Journal*.[220] As Moody's Investor Service warned creditors, the state faces a "severe loss of working-age residents due to long-term demographic trends and the state's susceptibility to natural disasters."[221] Moody's concluded that "expensive insurance has the potential to contribute to out-migration."[222] Carolyn Kousky, the leading scholar on climate change and insurance, noted that "insurance is where many people are feeling the economic impacts of climate change first. . . . That is going to spill over into housing markets, mortgage markets, and local economies."[223] Going a step further, Federal Reserve Governor Christopher Waller—a Trump appointee—acknowledged that "it is possible some of these physical [climate] risks could contribute to an exodus of people from certain cities or regions."[224]

For many climigrants and relocated households, it could very well be the costs and availability of insurance that contributed to their propensity to move. For others who are less fortunate, it will be the flood, fire, or hurricane that they were not insured for that will be the threshold factor driving their relocation. Research in economic history uncovered that in oil-dependent geographies in the American South, like southeast Louisiana, church membership is higher because it acted as a form of social insurance that "limits spillover [effects] of oil price shocks."[225] One spillover effect of oil is climate change. Unfortunately, many people are left each hurricane season with nothing more than a prayer that the next hurricane will not be "the one" that changes their lives forever.

The Climate Intelligence Arms Race

The increased costs of climate change are significant. As map 4.1 highlights, the costs associated with mortality, energy, and local GDP are projected to significantly burden much of the United States, with significant losses accruing in the Sun Belt. The refrain in the free market rhetoric that dominates the economic policies of many Southern states like Florida and Louisiana is "listen to the market." Unfortunately, many do not want to hear what the market has to say. For the financial services industry, they not only want to listen—they are in a race to hear the message before anyone else does. They want to shape the landscape for future regulations, and they want to use information asymmetries in the marketplace to their competitive economic advantage. They want to make money from climate change.

Across the professional services that support nearly every sector and asset class, there is a *climate intelligence arms race* to understand the emerging nature of physical and transition risks.[226] Everyone from rating agencies to venture capital firms are pouring money into companies and start-ups that are selling the hottest new climate intelligence products. Many firms

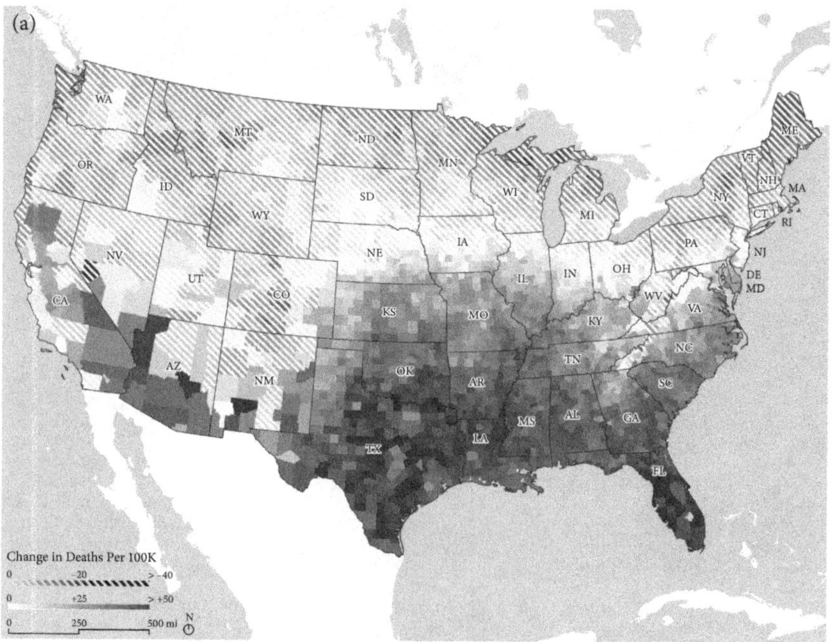

Map 4.1 Projected Human and Economic Impacts of Living in the Sun Belt Under RCP 8.5 (2080–2099): (a) Mortality; (b) Energy Expenditures; (c) Total Direct Damages. Design Credit: Oliver Oglesby and Jesse M. Keenan. Data Source: Hsiang, Solomon, Robert Kopp, Amir Jina, James Rising, Michael Delgado, Shashank Mohan, Daniel J. Rasmussen, Robert Muir-Wood, Paul Wilson, Michael Oppenheimer, Kate Larsen, and Trevor Houser. 2017. "Estimating economic damage from climate change in the United States." *Science* 356 (6345): 1362–1369. https://doi.org/10.1126/science.aal4369. Note: (a) and (b) contain 2022 updated data from ESRI.

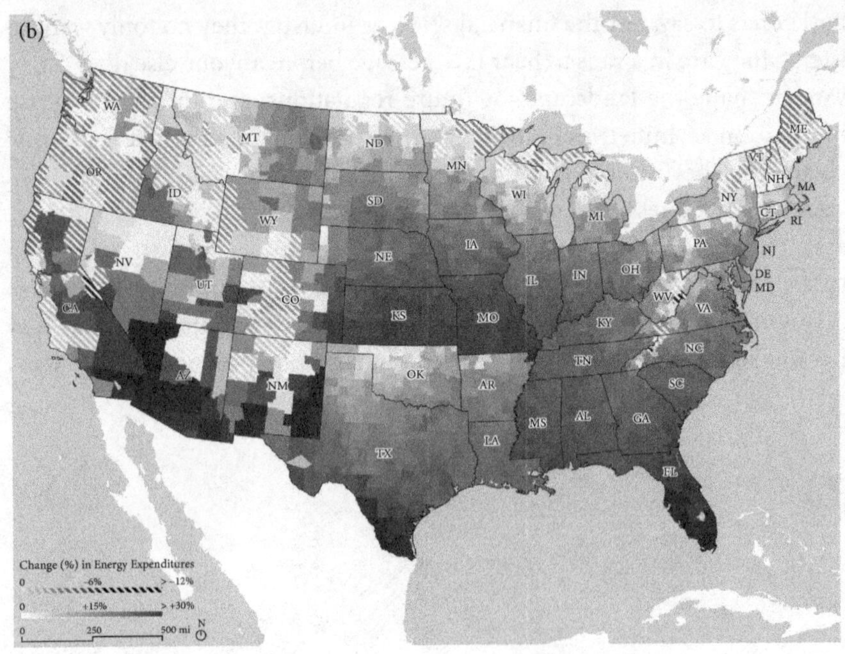

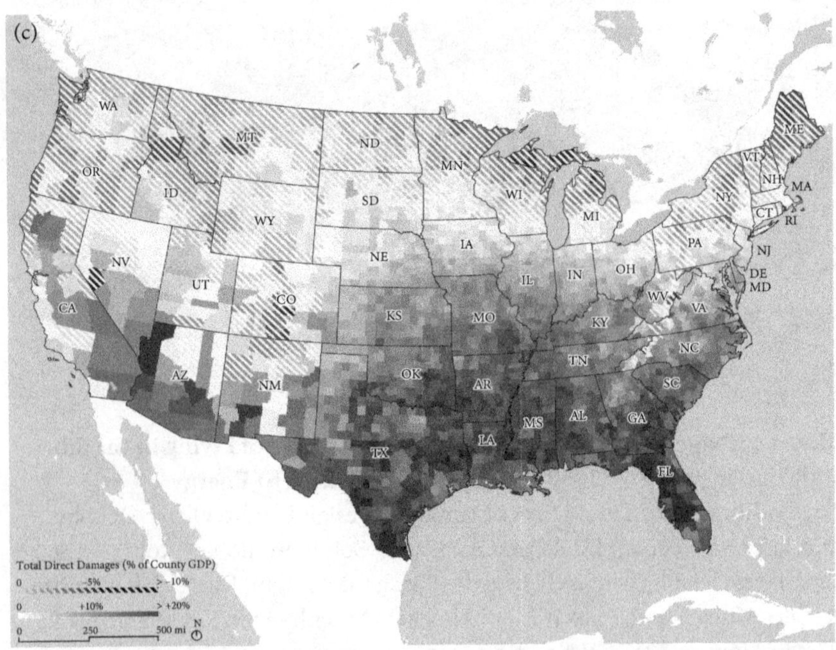

Map 4.1 Continued

are marketing such products, but the reality is that behind each product is a service that customizes an analysis for each use case and customer. Some of these products and services are built on the backs of academics who have labored for decades to refine something of value. Others are using their AI wizardry to sell unsophisticated customers experimental models with sufficient errors and uncertainties that one might regard these firms as mere hucksters. Many firms proclaim their mission to support consumer awareness without acknowledging their core strategy to offer free products to build a name brand that will yield future paying customers. At the same time, consumers and governments are often relying on unvalidated private science to make major decisions without appreciating the uncertainties behind these decision-support tools.

Between 2016 and 2019, Google and other tech companies sought to build a virtual planet where open-source code and streams of public and private climate, weather, impact, exposure, and even carbon data merge seamlessly into a single user experience. Google even assigned X, the moonshot factory, to the task of designing this virtual planet. Interface design with climate information had come a long way in the previous decade, and engineers and designers were convinced that they could design data layers with varying opacities within axonometric projections that could peel off a spinning virtual planet. Their mission was defined by the dual ambition to promote consumer engagement and to profit from the data integration services. They soon realized the science could not support the totality of their ambitions, so they elected to focus on water and flooding. After less than a year, they gave up. Not only was the science not sufficiently developed, but experts also warned them that the science may never be sufficiently developed for such an undertaking—at least not in their lifetimes.

Other climate services technology firms sometimes believe that advances in computation and technology can leap beyond the primary science. As the markets internalize proprietary private climate science, they create a number of fundamental problems. First, when this science is developed outside of the boundaries of double-blind peer review and other scientific consensus processes, the underlying methodologies and findings are not sufficiently scrutinized for their flaws and limitations. This is bad for science. When researchers publish this private climate science in scientific journals, they are often only publishing limited data outputs and are not subjecting their methodologies to review. This compromises scientific integrity. As some scholars have noted, some of these firms may be "operating outside of the bounds of scientific merit leading to the potential misuse and abuse of climate models."[227]

Many other firms are making strides in the responsible commercialization of science. Assuming that this proprietary private science does offer some value, it sets up the second problem. The marketplace is increasingly being bifurcated into public and private streams of information. As Dartmouth's Justin Mankin put it:

> I regard climate information as a public good and fear contributing to a world in which information about the unfolding risks of droughts, floods, wildfires, extreme heat and rising seas are hidden behind paywalls. People and companies who can afford private risk assessments will rent, buy and establish homes and businesses in safer places than the billions of others who can't, compounding disadvantage and leaving the most vulnerable among us exposed.[228]

When consumers in high-risk areas struggle to find reliable climate risk information, this runs the risk that consumers become complacent in taking the steps necessary to adapt. Worse, there is a scenario where the marketplace profits off their complacency. For instance, investors can short mortgage defaults or regional economic producers after an extreme event. Whether it is delayed information, misinformation, or no information at all, the resulting impacts on consumers are likely to result in everything from disorderly relocations to outright exploitation.

Governments want to promote the research and development necessary to drive advances in private science and technology, but they also want to ensure that these advances do not come at the cost of declines in social welfare in high-risk areas, including sending zones. According to one scenario by the legal scholar and engineer Madison Condon:

> Municipalities that want to challenge insurance and bond rating determinations must rally significant resources for modeling and data, a scattershot policing method at best. And when companies have access to sophisticated modeling about future impacts—some of them potentially devastating for entire communities—the decision to share that information has been largely left up to the corporation.[229]

In this scenario, based on real-world experiences, local governments reliant on private-sector climate analytics are limited in their capacity to warn their residents about the risks they face. At the end of the day, it is not uncommon for a local government to rely on the private sector to dictate the terms of tough decisions, because it partially insulates them from political culpability among those who are on the losing end of the equation. To make matters more challenging, local governments are sometimes disincentivized from publicly sharing information about their climate vulnerabilities because they may pay a higher price in the muni bond market for such disclosures. Paying a higher price in the bond market may ultimately result in fewer financial resources for infrastructure services and adaptation investments.

Free and accessible climate information is key to protecting people in harm's way. It is also key to shaping market forces. At the same time, many barriers exist that challenge the validity of price signal theory to shape market behavior. As outlined in this chapter, local beliefs about present and future climate risks shape everything from housing values to insurance market participation. Consumers should be free to hear what the market is telling them. Despite the U.S. Securities and Exchange Commission (SEC) caving to pressure from the second Trump administration to kill its climate disclosure rule, global markets are rapidly advancing disclosures drawn from the Task Force on Climate-Related Financial Disclosure (TCFD) guidance. Markets will soon be supercharged with information that will accelerate many of the discounting, revaluation, and other climate-informed behaviors outlined in this chapter. The climate intelligence arms race runs the risk of leaving many people behind who cannot afford to keep up.

By steering capital that has internalized the risks of climate change, the private sector is firmly in control of dictating the terms of future adaptation investments for consumers and for the public sector. Yielding this power to the private sector runs the risks of yielding a broader legitimacy of the public sector to protect its citizens and residents, particularly in high-risk sending zones. As people lose faith in the power and capacity of local governments to address climate change, they lose faith in the prospects of future prosperity in their present condition. It is this hopelessness that may very well break the bonds of place and community in a broader market-driven exodus of people.

5
Sending Zones, Climate Gentrification, and Unwinding Towns

Between physical climate impacts, a lack of public sector adaptive capacity, and the rapid adaptation of market economies, there is a wide array of conditions that may operate to "push" people and firms to migrate and relocate. There are also "pull" factors that draw people in with the promise of a superior way of life with less risk and lower costs. This chapter explores the push and pull of a wide range of climigration and relocation actors within sending zones. This chapter focuses on everything from the mechanics of how people dissociate from the status quo to how local governments begin to unwind in the face of a dwindling population and tax base. Sometimes there is a change in demand preferences that drives dislocation from climate gentrification, and, in other cases, there is a gradual decline of public services that accelerates the decline of public order. As sending zones come into focus, some people will relocate, and others will be trapped. This represents a major challenge for local governments who must transition populations, infrastructure, and a tax base, while also supporting the immobile populations that slip through the cracks. This chapter concludes with an exploration of what is lost and what gets left behind along the way.

Defining Geographies and Actors

The prospects of moving north from sending zones is both a literal reference to a cardinal direction and a metaphor for the broader movements of people, economies, and ecologies that defy directional simplification. As represented in Figure 5.1, some geographies may be defined by their magnetic pull, while others may be defined by the overwhelming weight of their push factors. While this binary is an incomplete representation of how people actually make relocation and migration decisions, there is an undeniable

North. Jesse M. Keenan, Oxford University Press. © Oxford University Press (2025).
DOI: 10.1093/9780197641644.003.0006

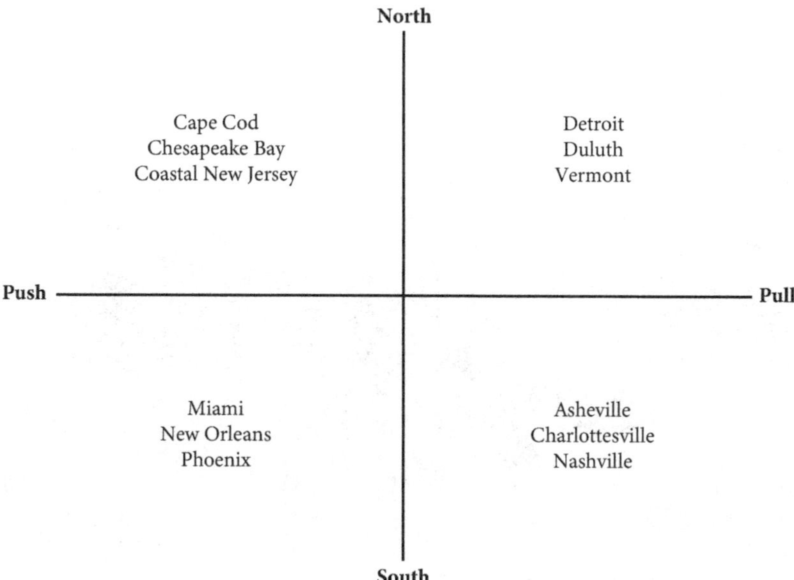

Figure 5.1 The Geographies of Push and Pull in the Face of Climate Change. Design Credit: Oliver Oglesby and Jesse M. Keenan

recognition that the decline of some places will result in the growth of other places.

For instance, Miami, New Orleans, and Phoenix are examples of places that likely lack the capacity for sustaining their current populations in the future. The seas will rise in Miami and New Orleans, and it will be dangerously hot without enough water in Phoenix. This is not even accounting for the intervening spread of disease, floods, heat-driven blackouts, and hurricanes. There will also be places in the Northeast that will see an exodus of people—albeit on a different scale. Coastal areas from Massachusetts to Maryland are already facing significant physical and economic stress from climate impacts. Sending zones exist throughout the country and are not entirely limited to the Sun Belt. That being said, the region is projected to see significant out-migration and relocations given its disproportionate exposure. From a big picture perspective highlighted in Map 5.1, much of the Sun Belt is projected to see significant declines in climatic suitability that have provided optimal conditions for human occupation over the past 6,000 years.

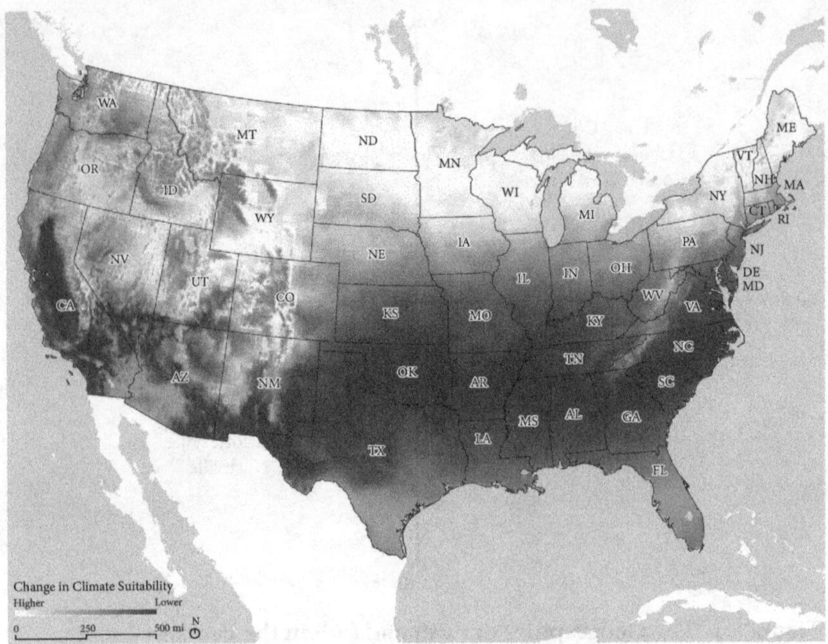

Map 5.1 Projected Change (Current–2070) in Climate Suitability and Temperature Niche for Humans. Design Credit: Oliver Oglesby and Jesse M. Keenan. Data Source: Xu, Chi, Timothy A. Kohler, Timothy M. Lenton, Jens-Christian Svenning, and Marten Scheffer. 2020. "Future of the human climate niche." *Proceedings of the National Academy of Sciences* 117(21): 11350–11355. https://doi.org/10.1073/pnas.1910114117

Receiving zones exist throughout the United States. As previously cited, many experts believe that most relocation activity will be local and regional in nature. However, some people are already moving across the country in the hopes of starting a new life in a place that is seemingly farther away from the ravages of climate change.[1] While these observed numbers of transcontinental movers are small, they highlight the emerging influence of climate on relocation behaviors.

As John Jenkins told *The New York Times* about his move to Duluth, "My decision wasn't based on just one thing, but climate was a big one. It's safer here."[2] Of course, nowhere is completely immune. What might look like a "climate haven" will have its own range of impacts. For Duluth, risks include lake level rise, extreme precipitation, and wildfires. After recent wildfire smoke inundation in Duluth, researcher Jamie Beck Alexander noted that "we moved here with a Pollyannish idea that we were moving to safety. But as someone who works in climate change, I know that the planet's systems are connected. You cannot pick up one part of the world and separate it from the rest of the planet."[3] While nowhere is safe, the push and pull of certain places is undeniable. Climate is rarely the sole motivation for moving, but it is increasingly among the top motivators. A 2023 survey conducted by *Forbes* magazine found that 30% of Americans were motivated to move, in part, by climate change.[4]

Not everyone who is relocating is fleeing climate change. Along a continuum of motivations outlined in Figure 5.2, there is a great deal of variation in different types of climigration and relocation actors. At one end of the continuum, there are people who are *episodically displaced* from extreme events. They lost their home and/or business, and they either cannot or do not want to rebuild. An experimental survey by the U.S. Census Bureau estimated that about 3.4 million people were displaced by natural disasters in 2022.[5] That is a little over 1% of the total U.S. population. Of this displaced population, over 500,000 people were estimated to be permanently displaced. This total current annual displaced population is approximately four times the proportionate annual share of the population that was displaced by the Dustbowl from 1930 to 1940.[6]

While the New Deal's Resettlement Administration, led by the Columbia economist Rexford Tugwell, never fulfilled its promise of building model resettlement towns due to the pollical headwinds of "socialist" planning,[7] it raises the prospects of whether any federal agency can navigate the local and national politics that plague moving people between receiving and sending zones.[8] Yet, government managed relocation is nothing new. Historians have documented dozens of cases of communities being relocated due to environmental hazards in the United States going back over 140 years.[9] Today,

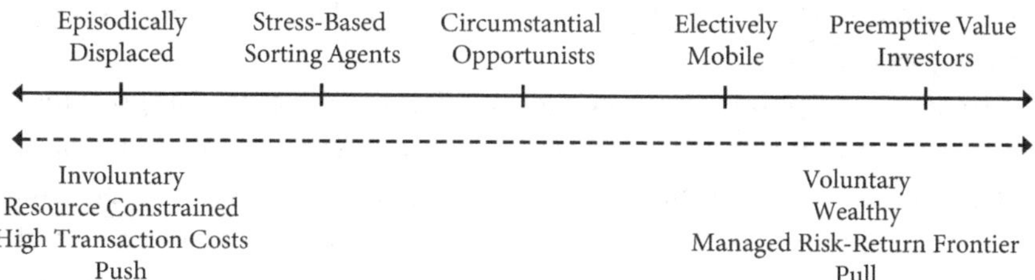

Figure 5.2 A Continuum of Relocation and Climigration Actors. Design Credit: Oliver Oglesby and Jesse M. Keenan

unfortunately, the federal government is largely agnostic as to where this 1% of the population goes every year.

For others who are not immediately displaced by extreme events, they may be sorting out of high-risk areas as they internalize the costs of climate impacts. As identified in chapter 4, these *stress-based sorting agents* are responding to the economic stresses by making trade-offs in where they are living based on the perceived risks and the costs of housing, utilities, insurance, and other financial services. Other actors may be defined as *circumstantial opportunists*. They may not be immediately concerned with climate impacts, and they may not be consciously making trade-offs between the costs and risks in alternative places. Yet, when presented with the opportunity for a lower cost of living with less risk, they jump at the chance. They may be seeking to make rational economic decisions, such as capturing the opportunity of owning a home in comparatively lower-risk area.

While circumstantial opportunists rank climate among a host of considerations, there are others who are primarily concerned about the present and future state of climate impacts on where they are currently living. This group of actors is not being pushed out by displacement or economic stress, rather they are making elective decisions to migrate or relocate in advance of future conditions. This group of *electively mobile* households are generally wealthier and have the labor skills that allow for a smoother transition across greater spans of geography. They may have accumulated wealth in a high-income state like California, or they have accrued household equity in fast growing states like Florida. They may work in white collar jobs that allow them to work remotely. This wealth and labor capacity allows them to shop and seek-out receiving zones that satisfy their preferences for everything from environmental amenities to the quality of school and medical systems. As the podcaster Greg Peterson noted about his move from Phoenix to Asheville, "We were looking for a place that was more rural, that received more rain, had better air quality and had a thriving local food growing scene."[10]

The electively mobile are related to a final group referenced as *preemptive value investors*. This group is actively seeking to invest in receiving zones in order to profit from the future influx of climigrants. These investors often focus on long-term land holdings and income producing apartment rentals. Their logic is that incoming climigrants will rent before they settle down and buy. Often the apartment income cross-subsidizes the holding costs of undeveloped land while they wait for the population influx to drive-up more demand over the long-term.

But, there are real limits to this strategy. As Owen Woolcock from Climate Core Capital noted, "We created the firm to give real estate investors a chance to be thoughtful about their portfolio allocations over the short and long-term," but "receiver communities may be reluctant to accept [development] or build fast enough" to keep up with the growth. As such, much of this activity is happening under the radar. Interviews conducted over a decade ago uncovered widespread land investments in central Florida made by patient foreign investors who anticipated a long-term exodus of South Floridians. Early-stage reports from real estate brokers in places like Duluth and Buffalo suggest a steady growth in transactions driven by outside investors. While exact figures are elusive, there is no doubt that people are preemptively investing in places poised to grow with climigration.

The Unwinding of Towns and Households

The various actors involved in relocation and climigration have a variety of different motivations. Some may be forced out, while others may be drawn in by the economic opportunities. Over time, local governments and households in sending zones may begin to "unwind" from the exodus of people and capital. This unwinding speaks to the internal mechanisms and feedbacks that may accelerate the exodus. For local governments, unwinding may simply relate to the gradual decline in the tax base and the provision of government services, while the unwinding of households relates to the gradual mental and physical transition from one place to another.

The individual and household motivations behind a sending zone exodus are complex and multi-faceted. The actors outlined in Figure 5.2 are composite groups, where the boundaries between each group exist along an undefined continuum of decision-making factors and preferences. While some scholars have sought to operationalize the decisions by individuals to relocate and policymakers to manage that relocation, there is no one-size fits all model for supporting decision making.[11] Sometimes the affective ambition to escape troubled circumstances defies rationality.

Research has only recently begun to understand relocation and climigration decision-making, with much of the focus on displaced persons.[12] As outlined in previous chapters, chronic physical impacts and economic stress are arguably just as likely to drive relocation as extreme events. To this end, Figure 5.3 explores the phases that everyone from the episodically displaced to the electively mobile might share in common when it comes to the gradual

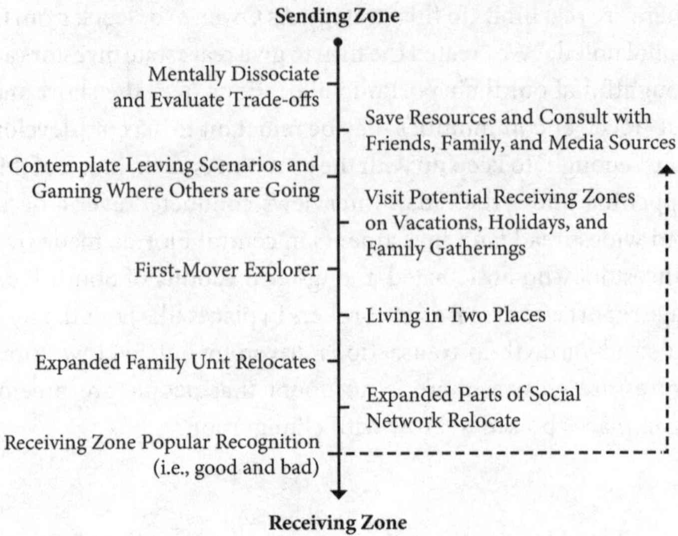

Figure 5.3 Potential Phases of Household Climate-Driven Relocation. Design Credit: Oliver Oglesby and Jesse M. Keenan

transition from one place to another. While this process is nonlinear and subject to intervening considerations, this simplified representation is useful for understanding the range of complexities that shape not just decision-making, but also the social diffusion of information about the fitness and desirability of receiving zones.

The first component of any transition from one place to another involves a gradual mental dissociation and emergent awareness of the trade-offs that define the status quo. Both unhealthy and healthy forms of mental dissociation may arise from a particular mental trauma from an extreme event[13] or as a means to cope with a changing environmental context within a broader understanding of psychological resilience.[14] The next phase is defined by actors' saving money and consulting with friends, family, and even media sources in order to both confirm that the sending zone is unfit and to identify a range of potential receiving zones.

Sometimes people, such as Southern California olive grower Zach Thorp, seek to review the science concerning the "the 'livability' of certain regions in the U.S." in the belief that there is "a prediction model or website that could be consulted" that would facilitate their decision-making.[15] For Thorp, he was not just interested in moving his family, he also sought to move his farming practices to a receiving zone that would be suitable for olive cultivation. While some scholars have attempted to develop notable models such as the Relocation Suitability Index, such decision-support tools are often intrinsically flawed in their underlying determinism.[16]

With adequate motivation and emerging resources, actors may then begin to construct different scenarios that represent tipping points for what would cause them to finally pack-up and leave. They are also prospectively developing mental scenarios that explore the fitness of different receiving zones. Interviews suggest that prospective climigrants may also be gaming where others are and are not going, with the fear that they may be caught-up in an exodus to a particular receiving zone where resources for handling such in-migration may be constrained. Actors may even visit their potential receiving zones on holidays or family gatherings. For instance, recent research has found that "Thanksgiving travel patterns correlate with relocations following Hurricanes Harvey, Irma, and Maria."[17]

Once a receiving zone is selected and it matches with a tipping point for leaving, individuals or households may make the plunge to be a first mover. For others, they may temporarily live in two places at the same time as a means to maintain and build social connections.[18] If the first-mover explorer climigrants are successful in the sending zones, they may pave the way for family members and others in their social network. Ultimately, the collective success or hardship of these moves provides a basis for the diffusion of social learning that creates a feedback loop for shaping the decisions of others. What households might not be able to control is the extent to which local governments may take actions that facilitate or accelerate their decision making.

For local governments in high-risk areas that face having or being future sending zones, there are two primary pathways—adaptive managed relocation or maladaptive disorderly decline. "Unwinding towns" are jurisdictions whose maladaptive policies accelerate what would otherwise have been the gradual out-migration or stabilization of its population. Table 5.1 provides a contextual pathway that may lead to this unwinding. It starts with policymakers sending mixed signals. While some might want to preemptively restrict activities in high-risk areas, others may seek to appease local political interests for maintaining the status quo.

The lack of a clear agenda almost always reinforces the status quo. As a mediation, local governments sometimes double down on engineering resilience investments in order to preserve the status quo. In this ambiguous

Table 5.1 Policy Playbook for Local Managed Relocation by Local Governments.

	Early-Stage Policies	Late-Stage Policies
Communications and Community Planning	Send policy signals of potential restrictions on development	Support post-buyout and relocation support for homeowners and renters
	Conduct outreach and plan for buyouts and community relocations	Conduct ongoing vulnerability assessments
Land Use and Zoning Policies	Actively down-zone in sending zones	Up-zone in receiving zones
	Assign non-conforming uses (e.g., landfills) in sending zones	Deploy TDR system between sending and receiving zones
	Assign inclusionary zoning in overlays in receiving zones	Shift property and economic tax base
Capital Planning Policies	Model life-cycle cost accounting of assets	Execute buyouts and community relocation
	Plan for decommissioning of assets	Deploy capital investments in infrastructure in local receiving zones

policy space, housing and commercial development continue to be built in high-risk areas, and local governments continue to lock themselves into long-term infrastructure obligations in order to support this development. As Valerie Nelson of the Cape Ann Coalition in Massachusetts noted, "We need . . . to understand the shortsighted trade-off in tax revenue for allowing high-risk coastal mega-mansions . . . against the long-term infrastructure costs for supporting an eroding coast."[19]

Local governments who fail to adequately assess their climate vulnerabilities set up one of several problems. First, they fail to communicate with local residents the risks that they face. This would have otherwise helped steer supply and demand away from high-risk areas, which would have come at a minimal cost to local governments—assuming that people and their tax dollars stay within the same jurisdiction. Second, without climate vulnerability assessments, local governments fail to properly invest in hazard mitigation and engineering resilience investments that could protect their tax base. Worse, they may simply seek to optimize their limited resources by protecting only the wealthy taxpayers.

Finally, this lack of foresight and planning is likely to result in a less than favorable credit rating when it comes time to issue muni bonds. The subsequent higher borrowing costs, together with increased operating and capital expenses from the climate impacts themselves, begin to become a fiscal strain that results in declines in public services. Similar to Columbia economist Stijn Van Nieuwerburgh's "urban doom loop" theory, the declining public services operate to accelerate the decline in the tax base, which, in turn, limits a local government's capacity to invest in resilience and adaptation.[20] The resulting *climate doom loop* further drives an exodus of people and capital resulting in the eventual unwinding of towns and local governments.

The climate doom loop highlights the potential risk of a local government finding themselves in an *adaptation growth paradox*. As the costs grow from climate change, they have to keep swimming, or they will sink. They have very little political or economic incentive to pull back from growing a tax base in the near-term. In part, they paradoxically need new tax-paying development, in often high-risk areas with lots of environmental amenities, in order to fund resilience and adaptation investments for existing parts of their jurisdiction that lack the infrastructural adaptive capacity to manage emerging climate impacts. This is one of the reasons that sophisticated local governments like Boston and Miami push for engineering resilience standards for new development in high-risk areas, they likely cannot afford to protect that area later on because they have already spent that future tax revenue elsewhere. Local officials often morally justify this house of cards by noting off-the-record that the high-risk new development—with often incomplete measures of engineering resilience—is mostly comprised of rich people who can afford the losses. The limits of this logic are rarely challenged in local public policy.

For jurisdictions with sending zones, Table 5.2 outlines a playbook for mitigating the worst of the climate doom loop scenario and for avoiding the adaptation paradox trap. A local government can assess and communicate the risks and send clear signals about emerging policies. Thereafter, the playbook is pretty simple. Local governments down-zone and restrict development in higher-risks areas and up-zone and increase development in lower-risks area. This effort is synchronized with an alignment of their capital plans for infrastructure investment that can decelerate private development in higher-risk areas (e.g., local sending zones) with asset decommissioning and accelerate private development in lower-risk areas (e.g., local receiving zones) with new infrastructure investments.

Table 5.2 Potential Phases of the Unwinding of Towns in a Climate Doom Loop.

Early-Stage Unwinding	Late-Stage Unwinding
Mixed policy signals	Incur greater project WACC and increased capital and O&M costs
Allow high-risk residential and commercial development	Raise taxes, fees, and utility rates
Fail to evaluate and communicate physical climate risks to public	Decrease public and infrastructure services
Underinvest in hazard mitigation to protect critical infrastructure and tax base	Decrease in economic competitiveness and affordability
Concentrate hazard mitigation efforts for wealthy only	Accelerated declines in property and economic tax base
Fail to account for asset impairment on the books	Loss of public confidence
Incur negative muni bond credit rating	Exodus of people and firms

These efforts may be done in tandem with individual buyouts and community relocation efforts, particularly because people participating in these processes benefit from well-resourced and planned receiving zones. While community relocation is a more expensive and heavier policy lift, it offers the significant opportunity to keep communities partially intact.[21] Unfortunately, the first major experiments in climate-driven community relocation in Newtok Village, Alaska, and Isle de Jean Charles, Louisiana, demonstrate how difficult the process can be. In Newtok, a splintering of federal agencies under the Biden administration grossly under-resourced the development of the receiving zone and that drove residents to split their lives between the sending and receiving zones.[22] In Isle de Jean Charles, internal conflicts over governance, a poor-quality receiving zone, and a lack of a spatial connection to the cultural landscape of the sending zone have operated to splinter an already precarious community.[23]

For many places that lack a strong community-driven consensus about when and where to go, individual buyouts may be the only remaining option. As referenced in chapter 3, many people participating in buyouts are left to their own devices when it comes to where they are going to live next. As such, many people simply move to another low-cost, high-risk area. Unfortunately, this playbook works best in jurisdictions that have a combination of intra-jurisdictional sending and receiving zones and are not existentially threatened by the prospects of an overwhelming climate doom loop (e.g., New Orleans, Charleston). For these jurisdictions, municipal bankruptcy and state receivership may be their only path forward. Unfortunately, municipal bankruptcy is centered on the reorganization of debt and not the liquidation of assets. It was never contemplated that some places may no longer be fiscally viable governance bodies without habitable territory to tax.

For example, small coastal water utilities around Wilmington, North Carolina, are increasingly under stress from increased O&M costs and declining ratepayers. Interviews suggest that the state is pressuring larger municipal systems to annex these smaller failing utilities. If they say yes, then they are likely onboarding more liabilities than assets. If they say no, then they run the risk that the state forces the annexation as part of a bankruptcy proceeding. In either event, their credit rating and capital reserves risk being compromised in the process. As the largest urban area in the region, who will bail out Wilmington's water utility when they begin to take on water?

Climate Gentrification

Even when local governments do everything right in steering the supply of housing and infrastructure development away from high-risk areas, they are up against changes in consumer demand that are potentially destabilizing to the local population itself. *Climate gentrification* is a process by which changes in demand drive people and investment out of high-risk areas and into lower-risk areas in a manner that crowds-out existing residents.[24] Sometimes that crowding-out is local, as people move from one side of town to another. In other cases, the destabilizing effects of climigration in receiving zones across the country may be a direct consequence of climate gentrification. In either case, climate gentrification is likely the byproduct of the growing market signals and increased costs outlined in chapter 4.

Conventional forms of gentrification are primarily centered on supply. Developers identify a district or neighborhood that has uncaptured value, and they increase the supply of new housing and commercial space through renovation and new development. This has positive and negative implications, but it has the potential to displace residents as rents go up. By contrast, climate gentrification is about demand. As people become more aware of the costs and impacts of climate change, they change their preferences in a manner that discounts high-risk properties and increases the value in lower-risk properties. This shift in preferences may drive increased demand in areas that can be destabilizing to the existing population. This is particularly true for areas that were often farther from environmental amenities (e.g., recreational beaches) where low-income residents historically settled.

Climate gentrification can occur in three ways.[25] The first pathway is called the *Superior Investment Pathway*. Here, people move away from high-risk areas and into lower-risk areas because they see the lower-risk areas as a superior economic investment. In other cases, living in a high-risk area is desirable for some people because of access to environmental amenities (e.g., living on a barrier island with coastal views). In the *Cost-Burden Pathway*, the costs of living in such a place continue to go up in a manner where only the wealthiest can afford to remain in place. In this scenario, the non-wealthy are forced to move to lower-risk areas in order to find a more affordable place to live. For example, as cited in chapter 4, they move off Miami Beach to the mainland in order to get more affordable car insurance coverage.

In some high-risk areas with sufficiently robust urban density, local governments are often inclined to make investments in engineering resilience and adaptation that seek to protect local residents. In the *Resilience Investment Pathway*, this investment has the unintended consequence of being treated as an additional amenity by the market. As the risks go down from the infrastructure-scaled risk reduction investments (or sometimes even a policy signal of future adaptation investments), then the value of protected properties goes up.[26] Often this resilience infrastructure includes appealing landscape designs that further increase the amenity value of adjacent properties.[27] As property values go up, the destabilization of local populations may follow.

As the demand for lower-risk properties heats up, existing residents in lower-risk areas face both a potential windfall for property owners and potential stress from increased rents for renters and small businesses. The Superior Investment Pathway has come to fruition in Miami and New Orleans, as demand for comparatively higher elevation land out of the flood zones has put development pressure on local working class, low income, and immigrant populations.[28] Sometimes this is not a gradual process but rather a sudden shock.

When more than 14,000 homes were destroyed in Paradise, California, in 2018, the neighboring lower-risk college town of Chico saw a "15–20% shift in population" almost overnight.[29] As scholars noted, "Chico's city services were stretched thin with additional traffic, degraded streets, an uptick in car accidents, robberies, and assaults, and a 16% increase in average daily sewage flows."[30] Within a matter of months, "home sales had doubled, and prices jumped 21%: Realtor.com even designated Chico the 'hottest market in the country.'"[31] For many former Paradise residents, Chico was not just a lifeline—it was a superior investment.

In coastal North Carolina, researchers have found evidence of housing disinvestment by the wealthy and the clustering of low-income and minority populations in high-risk areas, suggesting that the Superior Investment Pathway may sometimes be limited to people who have the resources to make investments in the first place.[32] Not everyone can afford to move. For those that can afford to move, they go to places like Asheville, North Carolina, in the Appalachian Mountains, which has seen significant development pressures accelerated by upper-income households moving off the coasts.[33] As one local interviewee noted, "I first noticed the shift in population when I picked my kids up at school and noticed lots of BMWs and Mercedes from car dealerships along the coast."[34] Accelerated by in-migration from California, Texas, and Florida, Asheville had the second highest in-to-out move ratio in the country in 2023.[35] For every resident that left, 3.29 residents moved in. Unfortunately, after the devastating flood from Hurricane Helene in 2024, people learned the hard lesson that nowhere is safe from climate change.

The Cost Burden Pathway has also come to fruition in various ways associated with post-disaster redevelopment. After repeated wildfires in Santa Rosa, California, only the wealthy could afford to rebuild high-performance fire resistant housing in high-risk areas.[36] In some cases, the new high performance post-disaster housing represented a doubling of the pre-disaster housing values—effectively crowding out many displaced residents.[37] This is consistent with research that has found that coastal adaptation investments themselves are strongly correlated with income and property values.[38] Adaptation has a cost and many cannot afford the expense.

After wildfires destroyed over 2,200 buildings and killed 102 people in Maui in 2023, the risk of climate gentrification was so great that the governor of Hawaii sought to issue a moratorium to prevent outsiders from coming in and buying land from families and rebuilding housing that was beyond the means of the displaced community members.[39] For Lahaina, adaptation interventions are centered on restoring native land management and water rights in order to reduce wildfire exposure, and these are expenses that many in the local community cannot afford. Owing to the convergence of colonial exploitation, economic stress, and environmental degradation, a majority of Native Hawaiians now live outside of Hawaii.[40]

Sometimes investments in resilience and adaptation have unintended consequences insofar as the uneven distribution of costs and benefits amplifies existing socioeconomic inequalities. Outside of Charleston, South Carolina, investments in green infrastructure along the coast to mitigate SLR have operated to increase land scarcity and accelerate displacement pressure on a historically Black community in manner consistent with the Resilience Investment Pathway.[41] Resilience investments in merely studying and mapping flood zones can also drive climate gentrification. For instance, a 2022 expansion of flood zones in the southside of Ithaca, New York—driven by changing patterns of extreme precipitation—increased the number of homes in a flood zone by 650%.[42] The resulting requirements to purchase flood insurance were observed to be a significant cost-burden for this diverse working-class neighborhood. As Cornell's Rebecca Brenner told *The Nation*, "We're expecting that when this flood map [is put into effect], it's gonna displace people. They're going to get forced from their homes."[43]

Ithaca highlights how wicked of a problem climate change really is. Seemingly far from the worst of climate change with abundant fresh water, Ithaca is probably a reasonable receiving zone for people leaving Long Island and coastal New Jersey. Yet, any city can have both local sending and receiving zones. There is no such thing as a climate haven, there are merely better and worse places to live. Ithaca is a progressive college town that is trying its best to do the right thing by studying and planning for future risks. In doing so, it creates an entirely new set of problems. For instance, when it finishes its flood wall in this newly expanded flood zone, that itself might drive climate gentrification burdens for some, while others benefit from increased housing investment in a newly minted lower-risk area. Adaptation for some may be maladaptation for others.

What can local governments do about climate gentrification? Using the playbook outlined in Table 5.1, they can gradually up-zone density in lower-risk places and impose inclusionary zoning requirements that require a certain percentage of new housing units to be affordable. They can reinforce the linkage between local sending and receiving zones with the use of transferable development rights (TDRs), wherein development capacity can be transferred from one place to another. For instance, in Norfolk, Virginia, a person in a high-risk flood zone can extinguish their development rights in exchange for development benefits in a lower-risk "Upland Resilience

Overlay district."[44] With a TDR system, local governments can even support community land trusts that offer the opportunity for the collective governance of new development that can preserve the affordability of housing. However, all these policy responses may also operate to accelerate climate gentrification by sending policy signals about disamenities and amenities in the respective sending and receiving zones. That may accelerate the destabilization of people who already live in these zones. As such, there may be no stopping climate gentrification. At best, local governments can merely delay the inevitable with the hopes of treating the casualties along the way.

What Gets Left Behind

Whether a sending zone represents a local relocation effort or a hotspot for driving transcontinental climigration, there are things that get left behind in the process. Sometimes people are left behind. One concern with climate gentrification is that as capital exits high-risk areas, low-income renters and even informal settlements fill the void, as is common in the Global South.[45] There are two competing perspectives concerning the future of America's coastal property markets—or any high-risk market with lots of environmental amenities for that matter. The first argument is that high-risk areas (e.g., barrier islands) will become increasingly wealthier pursuant to the Cost Burden Pathway. This increase in wealth crowds out investors in multi-family apartment buildings who will exit the market, and low-income renters will follow the shift in rental inventory somewhere else.[46]

The second argument is that "if these markets decline (e.g., outside property values adjust downward in response to climate risks, macroeconomic shocks, or tax reform), the result is an influx of lower income renters" because the value of the remaining properties is in today's rents and not tomorrow's future value.[47] This argument is consistent with the fact that many apartment investors are mostly concerned with rental income and they are not terribly concerned with the exit value of the property when they sell it. Some conservative investors in high-risk areas even model a zero-dollar exit value in their financial proformas.

The reality is that both arguments can be true at the same time, as "whether the coastal environment becomes low-income homogeneous, is populated by wealthy owner occupants, or is something in the middle" is dependent on "on policy interventions and the time paths of changes in the physical system and competing property markets."[48] In other words, anything can happen and the most that any local government can do is to be transparent about the risks, facilitate the transition, and treat the casualties through existing mechanisms of social welfare support.

In facilitating this transition, local governments with sending zones face an enormous challenge with managing what gets left behind. Some might called this "managed retreat" and others might called this "strategic relocation," but, ultimately, local governments are merely seeking to avoid maladaptation to something that is well underway by forces that are far beyond their control.[49] As chapter 4 argued, much of these forces are market forces that fundamentally undermine the democratic capacity of local governments to determine who gets helps and who is left to their own devices.

The challenge is enormous. Governments need to ensure that inundated and decommissioned properties, infrastructure, landfills, and toxic waste sites do not pollute the environment.[50] They need to document cultural and historically important places and practices,[51] and they need to preserve valuable archives.[52] This may even include documenting local recipes with local ingredients to facilitate the nutritional transition of climigrants.[53] Documenting local accents and language will also be important, as regional variation in language exists throughout America. Without preservation, the unique "American Elizabethan" accent of Sea Island residents in Chesapeake Bay will soon be lost to SLR.[54] Local governments may even need to leave historical monuments of what was left behind, particularly as this may be a powerful social incentive for continued environmental stewardship of sending zones.[55] As environmental stewards, they may even attempt to manage the relocation of species of flora and fauna through assisted colonization and migration.[56] Although, some have argued that this is a game of "ecological roulette."[57]

Local governments may find themselves with the broader obligation of supporting the development of peoples' livelihoods in their new place of residence.[58] By extension, they may need to help people move professional and business licenses and connect to new supply chains. They may very well need to digitize their public records so that future climigrants can access critical records for everything from education attainment to vaccinations. Older people will need help finding new healthcare providers and moving health records. Some climigrants may need new wardrobes (e.g., coats and wool socks) and personal property (e.g., space heaters) that helps them transition to a new climate in northern receiving zones. The needs are endless.

Researchers and policymakers are only now beginning to explore the wide range of challenges facing the support of people on the move. It is not just the living who will need to be relocated. Cemeteries at risk of inundation and destruction will need to be moved.[59] Indeed, cemeteries in receiving zones may very well need to be relocated to make room for the living. Brooklyn, New York, is one of the most desirable places to live in the United Stated and more than 10% of its land area contains cemeteries.[60] At a cost of up to $10,000 per gravesite, FEMA has already spent millions moving graves out of harms' way throughout the country.[61] That is a steep price that many of the living cannot even afford. Some national cemetery operators in Florida are already pitching the prospects of reinterning the deceased withing their network to prospective customers. It is never too early to plan for relocation and climigration—even after one dies.

6

Receiving Zones, Climate Havens, and Post-Climate America

History is littered with utopian visions for a better world. The world is also populated by utopian experiments—many of which are not so obviously successful. Driven mostly by gravity, the elevator is one of the most sustainable forms of transportation there is. It entirely liberated people from the calorie-burning limitations of urban density that had endured for millennia. It brought the world closer together—closer to doctors, closer to voting polls, and closer to loved ones. These experiments have come in all sizes from architectural design innovation to the broad strokes of social organization. The success and failure of these experiments is not just a measure of technological progress. It is the making and unmaking of the world that speaks to the very heart of what it means to adapt.[1]

This chapter advances the idea that receiving zones represent America's best shot at redefining a low-carbon, sustainable way of life. This is not a utopian idea. The tools, technologies, materials, and designs of sustainable progress already exist within widely developed supply chains of production. In this light, the renewable energy transition is simply a part of a broader transformation of the American way of life that seeks to reconcile its unbounded consumption. As with all forms of transformative adaptation, there are trade-offs. There are empirical limits to environmental sustainability and political limits to social change. There are also costs along the way that challenge the capacity of existing institutions, the self-governance of social organizations, and the power of the consumer. This chapter seeks to understand what one has to lose and gain in the search for a sustainable future in receiving zones.

This chapter begins with an exploration of the emerging discourse around the idea that certain receiving zones may be ideal locations that serve as a beacon—or "climate haven"—for future climigrants. This exploration evaluates the negative and positive implications of this popular utopian fiction.

North. Jesse M. Keenan, Oxford University Press. © Oxford University Press (2025).
DOI: 10.1093/9780197641644.003.0007

The chapter then shifts to outline some of the major elements of sustainable design and planning that may guide the development, renovation, and preservation of receiving zones and climate havens, alike. These tangible and often well-tested designs and plans are positioned within a range of observable indicators that define various forms of sustainability. Finally, the chapter concludes with the identification of pathways that might very well define a post-climate future in America. As global change drivers converge to shape how and where people will live and work, the optimism inherent in the drive for a sustainable future is tempered only by a lack of imagination.

The Promise and Peril of Climate Havens

In the fall of 2018, the widely respected environmental journalist Oliver Milman published a story in *The Guardian* titled, "Where Should You Move to Save Yourself from Climate Change?"[2] Driven by commentary from Portland State's Vivek Shandas and the author of this book, this story laid out the basic case for the identification of potential sending and receiving zones. The article argued that receiving zones will be defined by "areas towards the north" that might also have "sources of energy production [that] are stable, [with] cooler climates and . . . access to plenty of fresh water."[3] In a rhetorical flourish, Milman dubbed these receiving zones as "safe havens." This article would go on to spark the imagination of journalists, researchers, policymakers, and the general public from around the world. The sweeping ideas of climatic suitability and infrastructural capacity would be reframed in the media as "climate havens."

While "amenity migration" research has long sought to understand the pull factors of migration, this body of research has sustained "a long-standing debate over the relative influence of environmental and economic factors [that] has been inconclusive."[4] In the context of climate change, much of the research has focused on climate impacts as a disamenity that pushes people away, as opposed to lower levels of comparative risk being an amenity that pulls people in. For instance, research has shown that "both experiencing a disaster-level wildfire and extreme heat in the prior year were associated with reduced [in-]migration."[5] A survey of more than 1,110 California residents after the 2020 wildfire seasons found that "roughly a third of [the] sample intended to move in the next 5 years, nearly a quarter of whom reported that wildfire and smoke impacted their migration decision at least a moderate amount. Prior negative outcomes (e.g., evacuating, losing property) were associated with intentions to migrate."[6]

At best, one could infer that the disamenities associated with climate risks push people to places with lower risks, but the interacting complexity of various positive amenities in receiving zones associated with everything from labor market participation to lifestyle fitness remains largely unexplored in the United States.[7] The climigration scholar Alex de Sherbinbin and colleagues suggested that "environmental amenities and risks may be among the factors that affect aspirations and capabilities—but in this framing they cannot be said to 'drive' migration."[8] In this sense, climate change may simply push people to migrate or relocate who were already predisposed to take such actions. Other researchers argue that it may simply come down to money. They argue that the tipping point for out-migration is economic damages from extreme events: For every "$1,000 dollar[s of] damage per capita [there] is [an] associated . . . increase in out-migration" of between 9% and 16%.[9] Climate impacts as a push factor might be coming into focus, but the pull factors are not well understood.

The lack of theoretical and empirical clarity around the role of amenities in pulling people to receiving zones has not stopped scholars and journalists from driving a public discourse on which places might be climate havens and what climate havens should look like. For journalists, this motivation is somewhere between the clickbait production of the "Top Places to Move"[10] and a legitimate reflection on a challenge that appears to be crystal clear in light of the lived experiences of the people and places that define their beat.[11]

Many people recognize that climate change is already influencing where and how people live. In recent years, a group of urban planners and designers known as the PLACE Initiative gathered to identify a range of potential receiving zones based on a combination of factors ranging from climate risk to the quality of urban form. Map 6.1 highlights a range of geographies and places that were ultimately ranked as desirable potential receiving zones. The map also includes cities that have been observed to be receiving zones. While the data and methods of the PLACE Initiative are unvalidated and perhaps less than scientific, their work highlights a valuable starting point grounded by the professional judgment of those who are on the frontlines.

As receiving zones have come into focus, cities like Milwaukee[12] and Buffalo[13] have actively marketed themselves as climate havens. Buffalo has the great tagline: "How Buffalo's Weather Is Going from Punchline to Lifeline."[14] The marketing might even be working. According to Zillow, Buffalo

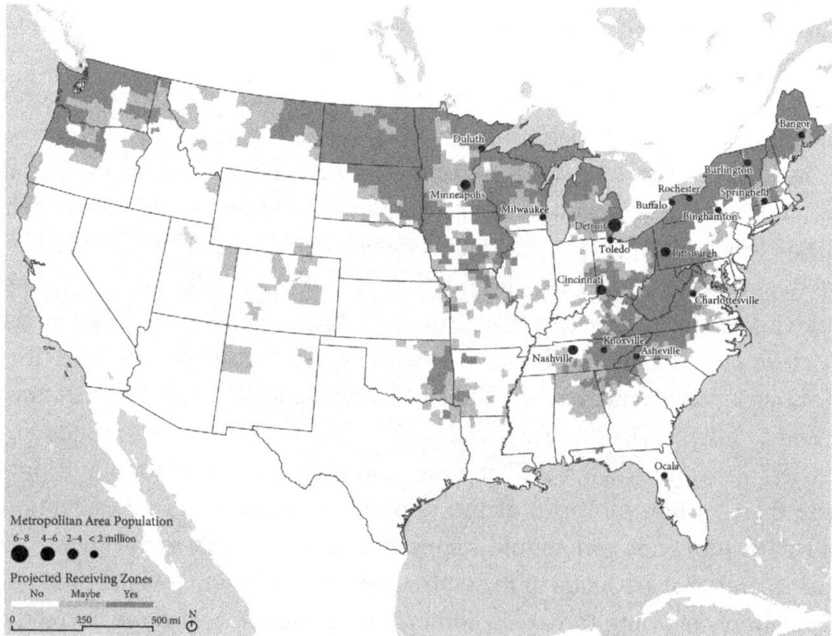

Map 6.1 PLACE Initiative Projected Receiving Zones with Observed Receiving Zones. Design Credit: Oliver Oglesby and Jesse M. Keenan. Data Source: PLACE Initiative. "Receiver Places." https://placeinitiative.org/receiver-places/.

has been the single hottest housing market in America from 2023 to 2025. These marketing efforts build on long-standing local policies to formalize welcoming efforts for immigrants—and by extension migrants—to the Midwest as a means to drive economic development.[15]

Commentators have raised both substantive and meritless challenges that highlight the promise and peril of the concept of climate havens. They argue that nowhere is safe and that no place can escape climate impacts.[16] This is very true. The flooding in Asheville, North Carolina, from Hurricane Helene in 2024 highlighted that even widely recognized receiving zones are still vulnerable to extreme events. Unfortunately, the history of post-disaster redevelopment in America suggests that, in a place like Asheville, the floods will likely be a catalyst for a post-development landscape that is spatially concentrated, built to a higher performance standard, and less affordable. It is likely that Hurricane Helene redevelopment will operate to both force people out and attract higher-income opportunists. At the end of the day, any receiving zone is vulnerable to extreme events. There is no refuge from a planetary crisis.

Other commentators have argued that a focus on climate havens ignores the plight of those left behind in sending zones. Some have even gone so far as to revive the long-dismissed binary of adaptation versus mitigation by questioning whether cities should prepare for climigrants or reduce their carbon footprint.[17] They argue that labeling some places a haven is misleading to potential climigrants, and that it is certainly not a haven for existing residents who are either currently cost burdened and under-served or might be crowded out through climate gentrification in the future.[18] Some tribal community members even see climigration as a kind of double colonization.

Some of these critiques are perfectly fair. Other critiques are grounded in baseless zero-sum rhetoric. First, cities can plan for climigration and mitigate their carbon footprint at the same time. Investments in adaptation and mitigation can and should happen in dialogue with each other. Any investments that are made in managing risk and carbon that benefit today's population are going to benefit tomorrow's population, if done correctly. For instance, investments in transit-oriented development (TOD) zoning and housing will reduce today's transit emissions,[19] but they will also provide a basis for future emissions reductions by driving greater measures of efficiency, walkability, and sustainability in dense mixed-income housing.[20] Mixed-income housing with lower transportation and energy costs will be key for supporting a diverse group of locals and climigrants.

Second, while no place can escape climate impacts, it is well established that impacts are unevenly distributed and concentrated in ways that define people's exposure and vulnerability.[21] Yes, there are extreme precipitation and wildfire risks in Vermont and upstate New York, but it is a relative picnic compared to what the Southwest and Southeast are facing, as was highlighted in chapter 2. Likewise, as outlined in chapters 3 and 4, the ideologically driven politics, anti-science belief systems, and widespread lack of institutional capacity in the Sun Belt operate to amplify these costs and vulnerabilities.

There is no denying that some places and people are comparatively better off. While many in the Northeast and the Rust Belt face legacies of economic exclusion, environmental injustice, generations of underinvestment, regional wildfires and droughts, and even scary tick-borne diseases, they are not facing the same existential convergence of risks that other regions face. Florida's future will no doubt be shaped by SLR, wildfires, salinification of groundwater, toxic and fecal contamination of drinking water systems, mega-hurricanes, stationary tropical systems, flooding of all types, extreme heat, and dengue and cholera outbreaks in ways that are almost unimaginable. As one commentator from Buffalo told the BBC, "We're not an oasis. We suck less."[22] There is definitely some truth in this statement, but it does not hold much weight for the thousands of Puerto Rican migrants who moved to Buffalo after the devastating 2017 hurricane season.[23]

Finally, the risk of misleading people through marketing a place as a kind of haven is low. As this book has highlighted, migration and relocation decisions are multifaceted and complex. A superficial marketing campaign might grab some attention, but it is unlikely to either drive large amounts of people to migrate or be a tipping point for those who might otherwise be predisposed. It is more likely that a city that both plans and markets for climigration will be successful in attracting climigrants. To this end, a more substantive critique is that cities engage in competitive economic development marketing but do not simultaneously advance investments in adaptation and climate mitigation planning. As the University at Buffalo's Nicholas Rajkovich noted, "In some cases, it's become more of an economic development slogan than the real detailed and robust planning that is going to be necessary."[24] This failure to substantively invest raises the prospects of crowding out and climate gentrification for the resident population that many commentators are concerned about. Climigrants may not only increase the cost of housing; they may also increase the costs of energy and other utilities that are anticipated to be strained under the weight of increased demand.[25] Planning for climigration will require cross-sectorial planning for everything from energy distribution to transportation.

Other commentators likely object to a potential scenario where the marketing of a climate haven draws in an economically mobile population that accelerates the hollowing out of sending zones, with increasingly high measures of immobile and impoverished people remaining behind. They object to the prospects of gaining from the loss of others. Again, there is no basis for believing that planning for climigration in receiving zones would substantively change the fate of sending zones that are destined to be depopulated. Giving people hope and a sense of optimism might very well outweigh the remote consequences of a negative self-perception. What commentators often fail to appreciate is that the audience for the marketing of climate havens is as much internal as it is external. It is about mobilizing the popular will to make investments in adaptation and mitigation by highlighting the real opportunities for sustainable growth. Although there are negative externalities to growth, sustainable development offers the only reasonable pathway for offsetting the costs and burdens of climate change.

The most damning substantive critique of climate havens was well captured by Juan Jhong-Chung from the Michigan Environmental Justice Coalition: "A lot of the disparities people are already facing are just going to be exacerbated by climate change. . . . It's a climate heaven for those that can afford it."[26] It takes money and resources to move, especially when it comes to the prospects of climigration from one part of the country to another. Many people simply do not have the resources. But mass migration events such as the one associated with Hurricane Katrina suggest that, while there is tremendous pain and loss in the sustained long-term out-migration in the subsequent years after an event, people of modest means have an enduring capacity to build news lives just about anywhere in the United States.[27] Hurricane Katrina demonstrated that migration connections to places receiving displaced persons actually reinforced relationships that helped support in-migration over the course of New Orlean's long recovery.[28] Receiving zones really can be lifelines in more ways than one.

The broader critique here suggests that there is a demand for state and local policy that can facilitate and subsidize low-income climigrants to ensure that people do not get left behind. In addition, local policies have to consider how climigration might drive crowding out and climate gentrification. These policies may include community investments in housing EE retrofits, renewable energy generation and storage, affordable housing production, inclusionary zoning, landlord-tenant alternative dispute resolution and legal representation, and other mechanisms that seek to reduce displacement. Again, these are investments that benefit current resident populations and also prepare for future climigrants who will be entering a higher-quality housing stock.

Climate havens might arise either naturally or by intent. But marketing and planning only go so far. Those places that offer the highest measures of social and environmental welfare for current residents are likely those that are going to be the most competitive for future residents. Different climate havens might arise for very different reasons. Some might offer a more diverse labor economy and strong public school systems that attract younger households. Other places might have robust medical systems and environmental leisure amenities that might attract older households. Some of these factors are within the ambit of local policies, and other factors might be a reflection of broader regional conditions that are largely independent of local and state policies.[29]

It may also be true that some climate havens might be unwilling to receive climigrants, and they may do everything in their power to drive exclusionary policies that make this abundantly clear. For those jurisdictions that do see the value of climigration, local governments can plan for climigration by using the same playbook that any jurisdiction would have for not only making investments in adaptation, mitigation, and sustainability but also investments in core issues associated with affordable housing, education and labor market training, public and mental health, social welfare, and transportation. For that matter, as outlined in chapter 5, even jurisdictions that are not going to be climate havens can abide by this playbook for helping guide the transition between local sending and receiving zones.

Right now, research suggests that climate havens are not ready for what comes next.[30] It is not just the physical limitations that define a receiving zone's carrying capacity; it is also their political, institutional, and cultural capacity to engage new residents. As Massachusetts resident and refugee resettlement coordinator Keegan Pyle noted:

> I am concerned about migration to the northern U.S. states due to climate change, and how the current residents will accept or challenge the change to their communities. I am concerned about housing, transportation, health care, and employment, but I don't hear anyone talking about the pushback that is bound to happen when more and more people are moving north.[31]

It is not just the crowding out of locals in receiving zones that may drive conflicts. Climigrants from the Deep South may bring their conservative political views to more progressive areas of the Northeast, and coastal liberals may bring their values to more conservative areas of the Midwest. While this potential phenomenon may create conflict, it might ultimately drive greater political diversity. For instance, it might work against the dubious theory of the "Big Sort,"[32] which suggests that America is increasingly sorting itself into homogenous clusters of like-minded people that fuels everything from gerrymandering to psychologically corrosive levels of social division.[33] Independent of one's theory of social sorting, climigration might very well drive positive social diversity with real political implications, particularly in Midwestern swing states.

The phraseology of climate havens might not be optimal given the competing and conflicting interpretations. But the language of refuges, safe harbors, havens, and utopias all fall into the same trap. This book utilizes the terminology of receiving zones in part because it implies a clinically neutral position on the complex trade-offs between environmental risk and physical and cultural carrying capacity. The phrase "climate destinations" might also be useful, as it balances interpretative meanings between geographies and amenities. Whether the terminology is climate destinations or climate havens, there is a substantive value in active language that gives people not only hope and optimism but also a sense of agency for dictating the terms of their future. It offers a psychological reprieve and a destination that may exist only as a mental model, and that is OK, too. People should be free to create their own refuge in full acknowledgement of the limitations of utopian thinking that often masks the true extent of the planetary crisis. It is not utopian to think that one can advance a sustainable future in a climate destination—even if that destination has its own serious social and environmental vulnerabilities. This is not climate opportunism. This is climate optimism.

Research in upstate New York suggests that are two dominant points of view on climate havens.[34] One perspective takes a "transformational view" that seeks to internally "address the social, political, and economic drivers of vulnerability."[35] The other point of view is largely external, with a focus on the "legal and legislative shortcomings in how to support climate migrants."[36] To be successful, local governments and communities are going to have acknowledge the validity of and the opportunities that come from both points of view. Whatever one calls it may or may not be so important. What matters is the difference that can be made in positively transforming peoples' lives.

Duluth, Minnesota

Domestic climigration to climate destinations has already transformed some peoples' lives. Duluth, Minnesota, represents an emerging case study that typifies the promise and peril of what it means to be a climate destination. In 2018, the author of this book was commissioned by the University of Minnesota Duluth to investigate how the city might plan for the growth that comes along with being a climate destination in the future. The subsequent research and 2019 public presentation were the basis of a documentary and a series of stories by *The New York Times*.[37] In the time since this research,[38] Duluth has been the focus of widespread public media attention, with documentaries and news stories from across the globe.[39] As a climate destination, Duluth was even the source for a comedic parody on *The Daily Show*[40] and a fictional plot point on Apple TV+.[41]

For many, Duluth might not be such a haven. It can be bitterly cold in the winter, with life-threatening levels of snowfall. There are the risks of wildfires, smoke inundation, flash floods, and lake level rise on the southern shores of Lake Superior. Duluth is a working-class city with a disproportionately high number of disabled persons, many of whose disabilities are likely the result of the material extraction industries that have historically driven their port economy. Coming of age at the end of the nineteenth century, Duluth has an aging housing stock, with many outstanding demands for historic preservation. With a peak population of 107,000 in 1960, the city's population experienced a precipitous decline to about 84,000 with the loss of heavy industries by the early 2000s.[42] As *The New York Times* noted, "The economic situation in Duluth was so grim that a billboard popped up on the highway leading out of town, bearing the message 'Will the last one leaving Duluth please turn out the light?'"[43]

Recognizing the writing on the wall, the State of Minnesota set out with a novel idea—what if Duluth could become a beacon of a new sustainable economy? The region has long relied on environmental extraction, and it possesses a great deal of local expertise at the intersection of resource management and heavy industry that could be transitioned to a new model for a green economy. People know how to make things, and they know how to limit their impact on the environment. With the help from the state, this drive toward a sustainable economy was reinforced by the region's cultural roots in ecocentric Scandinavia and even by progressive Lutherans who served as local sustainability leaders.[44] The region also benefited from the values, Indigenous knowledge, and environmental advocacy and stewardship of tribal communities such as the Fond du Lac Band of Lake Superior Chippewa. Local tribal leaders such as Karen Diver are regarded as some of the leading climate advocates in the country. Although it took time for the State's economic development vision to come together, it gave locals a powerful vision of what their future could look like. It was not just an economic development strategy; it was the creation of a new identity built around shared values.

Aside from an emerging green economy, Duluth also has the benefit of being a beautiful and affordable place to live, with good-quality public schools and a robust healthcare system that serves people as far away as the Dakotas. Perched on a hillside overlooking Lake Superior, the views offer a sublime reflection on the vastness of nature. In light of the emerging domestic and international scarcity of fresh water, Duluth is also sitting on a gold mine. Lake Superior contains 10% of the world's fresh surface water.[45] The city was planned for a much larger population, and it contains a diverse range of architectural styles and housing typologies. With a state university and medical school, an emerging culinary scene, and a postindustrial district for artists and craft breweries, Duluth is one of America's greatest untapped places to live—with or without climate change.

When the author of this book took on the Duluth research, the sales pitch was easy.[46] Although there were projected increases in a variety of impacts and climatic hazards, its climate was also projected to moderate in temperature, particularly in the winters. From its urban design to its affordable housing and from its welcoming cultural values to its recreational amenities, Duluth offered a great deal of opportunity for everyone, including climigrants.

The research in Duluth was undertaken on several fronts. First, as represented in Figure 6.1, small increases in population were spatially and

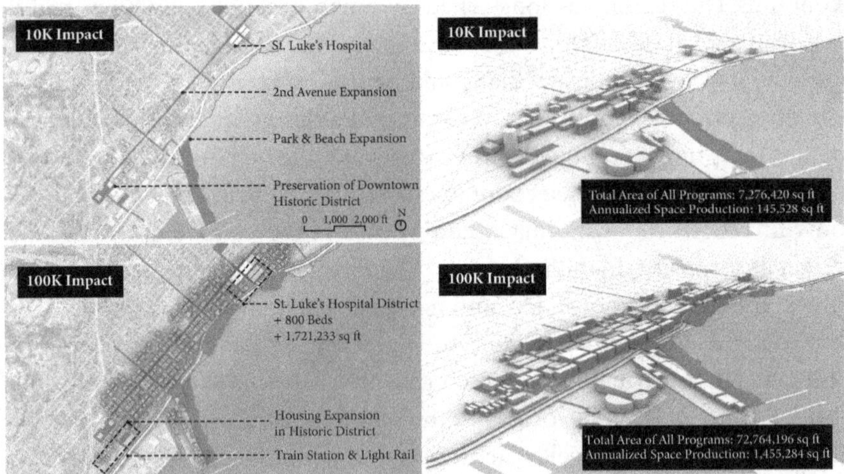

Figure 6.1 Urban Simulations of Incremental Climigration in Duluth, Minnesota (2019). Design Credit: Jesse M. Keenan and Harvard Graduate School of Design Duluth Research Team.

economically simulated over time to demonstrate the physical and fiscal impacts of climigrants. This simulation was centered on the proposition that climigrants should live downtown to avoid carbon-intensive sprawl. This exercise offered Duluthians the ability to understand the opportunities for sustainable urban development that were latent in the city design and infrastructure, as well as to understand the economic implications associated with taxing and paying for these investments. These simulations showed that small numbers of climigrants could have an outsized positive impact in driving urban transformation.

Second, the research identified sending zones throughout the country that were identified in climate demography research to understand who prospectively might be on the move. The project used the tools of marketing research to identify specific potential climigrant cohorts, including retired couples from Florida and working-age families from coastal Texas. Thereafter, a mock economic development marketing campaign was designed to test what ideas, slogans, and environmental imagery (e.g., amenities) connected with different potential climigrant cohorts.[47] In this sense, the cohorts were closer to the electively mobile than the episodically displaced. With these cohorts identified, social media tests were run with different marketing media to see what was effective. People in Florida definitely connected with visual cues of Lake Superior that mimicked the environmental amenities of coastal living.

Some found this exploratory methodology to be leading in a dangerous direction insofar as it would allow cities to cherry-pick wealthy demographics that would accelerate climate gentrification. That was not the intention. The intention was to demonstrate a potential tool for reaching a great diversity of people. Management consultants and crypto miners do not build homes and teach grade school students. In the age of AI, their power over the economy will wane in favor of those who build, make, and contribute to the things that society needs. Duluth needs these tools to attract healthcare and childcare professionals—not wealthy retirees.

The research offered Duluthians the opportunity to organize themselves in preparation for what would be, in the long term, an inevitable influx of climigrants. It was up to them to define who they were, how socially inclusive they wanted to be, and what kind of adaptation strategy would be best suited to their collective goals and values. For the most part, Duluth never actively branded itself as a climate haven. They did not have to. The global media did that for them. The general local consensus was aligned with a certain humility and an ambition to quietly advance their stewardship of the environment and to invest in upgrading their infrastructure.

In the years following this research, homes prices went up faster in Duluth than the national average, and rents rose at nearly double the national average, with the rental vacancy falling to just 2%.[48] An heir to the Cargill fortune even came into town and bought 10 homes in one neighborhood, stoking fears of climate gentrification.[49] Like much of the United States, Duluth has ridden a wave of economic prosperity in recent years that has increased median incomes and lowered unemployment rates. For generations, census estimates identified a statistically flat population with average net annual increases of around 50 people per year. Since 2019, Duluth's population has made a leap forward. More than 2,500 people have moved to Duluth in five years following this research.[50]

This increase in population has attracted more state and federal funding, but it has come at a cost. Rapid increases in housing costs have undermined everything from workforce development to childcare. As one local put it, "People from California, Utah, and places that are seeing climate change more dramatically in their area, [are] buying houses in Duluth sight-unseen. [They are] able to offer more than the asking price, with no contingencies, in cash, because the houses where they live are so expensive."[51] There is still time for Duluth to plan for climigration before it is too late. At this rate, they will have their own sending and receiving zones to worry about as people become increasingly crowded out and cost burdened. They were warned. While this might seem more like a self-fulling prophecy than a substantive trend to some, these critics are likely not listening closely enough to what people, markets, and governments are trying to say.

A Sustainable Future

Like adaptation, sustainability is a messy business. The call for climate destinations and receiving zones to be sustainable should not hide the tensions and conflicts that often mask what defines success. The measures of successful sustainable development could vary a great deal depending on the starting and ending points of people, communities, and planning jurisdictions. While there are objective measures of environmental sustainability in the built environment and in household consumption, the concept of sustainability is more or less a guiding set of principles rather than fixed benchmarks for absolute attainment. To this end, environmental sustainability is shaped by dimensions of both science and ethics.

Although a survey of sustainability science is well beyond the capacity of this chapter, the basic concept can be distilled in one of several forms. The best-known form was memorialized in the 1987 *Brundtland Report*, which suggests that *sustainability* is an exercise in human and economic development "that meets the needs of the present without jeopardizing the ability of future generations to meet their own needs."[52] This perspective centers on the idea of preserving resources to advance intergenerational equity. This perspective would later be operationalized by the concept of the "triple-bottom line," which imposes a consideration by economic agents to balance social, economic, and environmental trade-offs.[53] By contrast, biologists and ecologists define sustainability as "the rates at which renewable resources could be extracted or damaged by pollution without threatening the underlying integrity of ecosystems."[54] The question here is how much can humans extract and develop without irrevocably destroying the ecological systems that they rely on?

These three perspectives highlight the tension between the anthropocentric and environmental ends of sustainability. From an anthropocentric point of view, environmental capital can be sacrificed, as long as it advances human welfare and there is a search for a renewable substitute whose rate of consumption is less than its natural rate of replenishment. This is often referenced as *weak sustainability*. From a non-anthropocentric point view, natural and human capital are not interchangeable, and natural capital should not be depleted at all, particularly because many forms of natural capital are nonrenewable resources. This is referenced as *strong sustainability*. Under the concept of strong sustainability, nonrenewable resource consumption must decline at a rate that is greater than the rate of the resource's depletion. Unfortunately, sustainability in the built environment can only ever drive weak sustainability because of the inherent consumption of land, materials, and nonrenewable energy. Even the consumption of renewable energy in the built environment requires the consumption of nonrenewable resources.

Sustainability is often challenged because it is frequently sold as a win-win concept that can support limitless consumption.[55] Perpetual and limitless consumption is an impossibility under the laws of physics. To the contrary, sustainability in one system often comes at the cost of instability in another system. The classic example relates to the conundrum of the sustainable consumption of wood products in the built environment. If all wood products in the United States were forced to operate under sustainable certification standards, then the price of wood would go up as supplies become constrained. In due course, markets would look for substitutes for cheaper non-sustainable wood in other countries, which would drive deforestation in places like Indonesia and Brazil.

Carrying forward this thought experiment, suppose then that all wood products around the world would then need to be sustainably certified. What would then happen? Markets would seek to substitute sustainable wood products with cheaper and more carbon-intensive building materials, such as bricks, cement, and steel. Ultimately, if all materials were to be magically sustainable, then the tension would exist between sustainability goals and increasing scarcity in a growing world that demands more of everything. Sustainability has its limits in the real world. For this reason, there is active debate on the merits of green growth versus degrowth, which is a debate that is likely to shape many receiving zones in the years to come.[56] Degrowth might very well be used by some to justify the exclusion of climigrants whose primary point of reference is scarcity and the maintenance of a prior unsustainable way of life.

These limits to sustainability do not mean that under conditions of weak sustainability, the production and management of the built environment cannot seek to be advanced. In this context, sustainability is about reducing carbon and chemical emissions, the ecological footprint of development, and the demand for and consumption of energy and water. To accomplish this task, sustainable interventions in design, planning, engineering, and management exist from the scale of nanomaterials to the scale of regions.[57]

The biggest opportunities for sustainability start with housing and buildings. This is not rocket science. There are two goals—reduce emissions and impacts from building materials and reduce emissions and impacts from operations. Tools exist to support lifecycle analysis (LCA) of the total carbon equivalency and other harmful impacts that arise from everything from material processing to the shipping of that material to the worksite. Tools such as NIST's Building for Environmental and Economic Sustainability (BEES) program allow users to put a price on these materials and to optimize a range of material selections that minimize the building's overall footprint for everything from climate emissions to IAQ and from eutrophication (i.e., nutrient accumulation that drives harmful algae blooms in water bodies) impacts to ecological toxicity.[58] With the help of building codes, LCA tools make it possible to systematically reduce the footprint of every material that goes into the built environment.

Energy and heat come into and out of buildings by way of radiant energy, convection, conduction, and evaporation driven by the ambient environmental conditions and the material construction of the building. This flow of energy into and out of a building is known as an energy balance (kWh/m^2). The goal of sustainable design is to reduce the passive energy balance of a building to the extent that the building either requires very little or no external energy at all to run the systems that support the operations of the building (e.g., lighting, thermal conditioning). The United States has made significant advances in sustainability driven largely by energy codes that have focused on reducing energy demand and increasing efficiency by upgrading lighting and HVAC systems; sealing up and insulating walls, ceilings, windows, doors, and foundations; and reducing the amount of air exchange that happens between the inside and outside of the building.[59] It is not new, expensive technologies that have made the difference; it is time-tested construction and renovation techniques.

Every year innovative environmental technologies are entering the supply chain, from mold-resistant interior materials to home energy storage. Although these new technologies and materials do sometimes increase marginal costs, the marketplace is rapidly reaching near parity between sustainable and non-sustainable design because the cost savings of reduced energy and risk are increasing the payback for marginal upfront investments.[60] As such, sustainability is not a luxury amenity in housing; it is a critical way to advance affordable housing through the reduction of energy, utility, and insurance cost burdens.[61]

The one technological question in housing that remains unanswered is whether housing will morph into a machine of distributed and networked energy generation and storage that not only offsets costs but also generates income for resident investors.[62] If this does happen, then the implications for low-income residents may be severe. They may not only be unable to afford the upfront capital costs, but they may also pay higher public utility charges because of a declining public utility rate base that is increasingly going offline. This potential phenomenon begins to blur the distinctions between housing and infrastructure, as more forms of distributed infrastructure, from energy generation to wastewater processing, scale down to districts, neighborhoods, and buildings.

At a conventional infrastructural scale, the goals are the same—reduce the demand and increase the efficient use of energy and water, while at the same time reducing overall emissions and harmful impacts. Between material and process efficiency, many of these advances are low-technology and low-cost modifications to current practice that decrease the costs of operations and capital investment and increase long-term asset value. Sustainable infrastructure is also increasingly integrating engineering processes with natural processes through the incorporation of ecological systems into NBS and infrastructure design.[63] From landscapes that capture, store, and clean water to the stewardship of urban forestry that reduces urban heat island effects and energy demand, there is a vast range of emerging practices and co-benefits in green, blue-green, and blue infrastructure that are driving EBA investments in the United States.[64] These strategies often offer the benefits of reducing both carbon and physical risk.

Engineering resilience and adaptation planning also reduce the impact on the environment to the extent that these processes manage physical risks that would otherwise result in material loss or degradation from the shocks and stresses of climate impacts. As referenced in chapter 3, engineering resilience offers the opportunity for buildings and infrastructure to absorb external shocks and to revert to a single-equilibrium measure of performance. This is key for keeping the taps flowing and the lights on. By contrast, adaptation planning often acknowledges the lack of sustainability defining the status quo and seeks to set standards and define spatial regulations that reduce risks and shift assets and populations to more suitable locations. It is not just how one builds—it is also important where one builds.

Whether it is EE retrofits for the preservation of affordable housing or large-scale investments in regional water infrastructure, the public sector plays a critical role in planning, financing, and steering these investments. The public sector can drive demand by being an anchor customer. It can bulk purchase equipment to reduce unit costs. It can set up funds to reduce the WACC for nonprofit and for-profit developers. At the end of the day, the public sector can lead by example in driving sustainability in their assets and practices. All these activities are key for climate destinations and receiving zones that seek to find opportunity in what might otherwise appear to be a disorderly process of climigration and local relocation.

Between orderly planning and disorderly market-driven forces, the prospects of the sustainable settlement of climigrants and relocated households may vary greatly across the American landscape.[65] Although markets are increasingly internalizing the costs of carbon and climate risks in a manner that should drive economic motivation for sustainable investment, this assumes that there is capital available to make such investments. Many people simply make do with cheaper, inferior, and less sustainable materials and designs. Figure 6.2 highlights a wide range of potential settlement models that could manifest in both existing places and under conditions of new development. As highlighted in chapter 5, "unwinding towns" represent the socioeconomic and structural decline of jurisdictions that have failed to take the steps necessary to fiscally and physically stabilize in a manner that transitions a population and a tax base. In the case of new development, climate destinations and jurisdictions with receiving zones that fail to recognize the signals outlined in this book face the prospects of more unsustainable suburban and exurban sprawl.

For those jurisdictions that do recognize the opportunity, both sustainable redevelopment and the prospects of planning "new towns" are on the horizon. But the development of new towns and settlements presents a novel range of opportunities and challenges. Harvard urban planning scholars Ann Forsyth and Richard Peiser argue that, "due to land prices and availability[,] such developments are likely to be in a suburban or small-town location, rather than in established parts of larger cities."[66] They argue that one advantage to new town development is that because nearly everyone is a recent transplant, it will increase "the possibilities of making new social

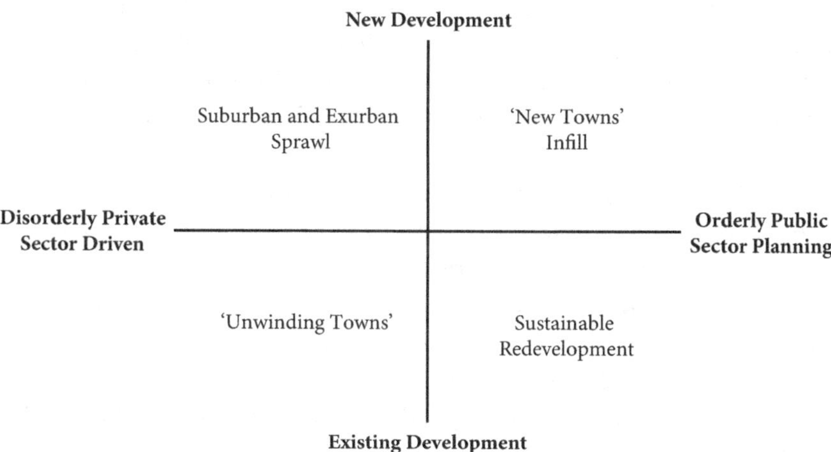

Figure 6.2 Relocation-Driven Settlement Typologies for Households and Communities. Design Credit: Oliver Oglesby and Jesse M. Keenan.

networks and community ties" because they are not burdened by existing strictures of social hierarchies and institutions.[67]

However, history has demonstrated that it requires a "heavy-handed" government to develop such new towns, wherein Forsyth and Peiser argue that "smaller new neighborhoods are easier to build and better match experience to date with whole community resettlement."[68] This might be true, but it raises a more fundamental question as to whether neighborhood-scale infill development—in the vein of new urbanism, for instance—can operate at a scale sufficient to support the types of infrastructure necessary for broader measures of sustainability and decarbonization. Whether it is new town development outside of metropolitan areas or infill development inside of metropolitan areas, there are difficult trade-offs associated with balancing existing carbon-intensive infrastructure systems with more expensive distributed and decentralized environmentally friendly infrastructure systems.

When it comes to infrastructure, scale matters and the costs are not trivial. Legacy infrastructure systems in infill areas might be more affordable in terms of the marginal cost of additional capacity, but they are difficult to decarbonize. The real emissions that matter most are land use, transportation, and building emissions, and these reductions might be more readily achieved through infill development with higher densities in metropolitan areas. Distributed sustainable systems that offer more technological innovation might be more desirable in new town non-metropolitan areas, but their benefits might be offset by much higher land use, transportation, and building emissions associated with larger homes on larger lots that emulate more recent suburban development typologies. In either case, without significant federal and state government interventions, the direct costs of utility rates and the imputed costs on land prices of new infrastructure may negatively shape the affordability and inclusivity of any new town or infill development absent other forms of subsidy. This raises the prospects for more exclusive forms of enclave development advanced by those wealthy enough to pay for their own sustainable infrastructure.

Worse, without federal and state intervention to support regional planning and growth management, new towns might very well drive long-term settlement patterns that negatively impact sensitive rural environments. Beyond the economies of scale for public infrastructure, moving to the middle of nowhere to develop a new town for climigrants is environmentally problematic. Some research has even identified a "climate-related counterurbanisation" phenomena where people seek to adapt to climate impacts by moving to rural areas without realizing the long-term maladaptive implications of their comparatively inefficient and ecologically disruptive footprint on rural environments.[69]

This is currently the story of Vermont, where some realtors report that upward of 25% of their clients are climigrants.[70] Climate-related counterurbanisation raises the prospects that the future of new town development should be optimally sited as infill development within existing urban and mature suburban areas that can support increased infrastructure demands at a lower unit cost, as well as amenity access for healthcare, education, jobs, transportation, and civic engagement. Unfortunately for Vermont, there is no regional plan for managing climigrants, and this risks further fragmenting its rural and urban gradient to the detriment of its agricultural economy and cultural way of life.[71]

As places like Vermont are starting to realize, the stakes are high. As represented in Figure 6.3, there is a scenario where environmental sustainability is sought by wealthy climigrants who live in exclusive enclaves that are packed full of high-tech environmental amenities with their own private infrastructural systems that allow them to disengage from the world around them—both physically and socially. The governance structures of these enclaves could range from more socially inclusive organizations, such as community trusts, to more economically exclusive organizations, such as housing cooperatives. A less than optimal scenario is that sustainability

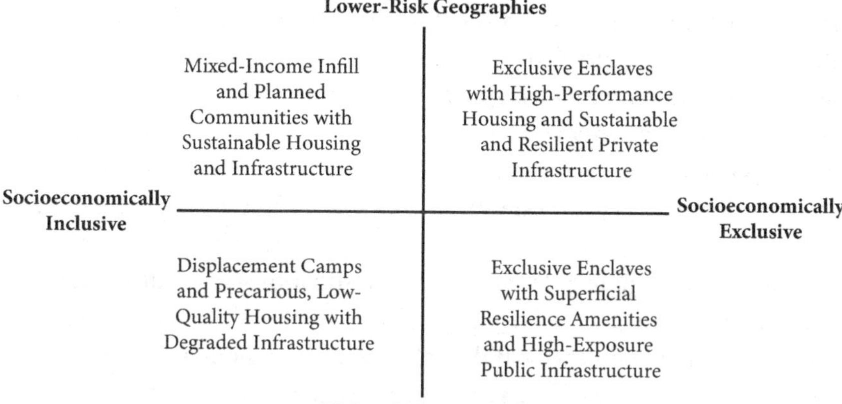

Figure 6.3 **Relocation-Driven Development Typologies for Households and Communities.** Design Credit: Oliver Oglesby and Jesse M. Keenan.

investments are isolated to those with the wealth to invest in decentralized technologies, and this allows them to seal themselves off in enclaves that mimic the worst of American suburban isolationism. This is what some critics of climate havens worry about.

Under a different and perhaps parallel scenario, non-wealthy climigrants get stuck in energy-intensive, poor-quality housing unprepared for increased energy costs and higher utility bills. As this chapter has outlined, it does not have to be this way. Both wealthy and non-wealthy households can advance sustainability through low-cost and low-tech investments that reduce impacts and costs, while at the same time finding a balance between private distributed district-scale sustainable infrastructure and connecting to public infrastructure that brings scaled efficiencies and regulated market dynamics. The developments can benefit from carbon- and risk-optimized zoning and land use that minimize emissions and physical risks. Of course, this is easier said than done in the context of greater scarcity of infrastructure-supported land, particularly as local sending zones expand.

As the renewable energy transition accelerates, these sustainable developments may spatially align with sources of renewable energy, particularly if long-distance transmission development efforts stall.[72] As highlighted by Map 6.2, inter-regional renewable energy zones (IREZ) for solar energy and onshore wind might support climigration receiving zones in the Midwest, while offshore wind might drive new development in New England.[73] IREZ-driven development may or may not work against the prospects of urban infill development. However, it might accelerate new forms of energy infrastructure that blend elements of housing and infrastructure. In one scenario, these IREZ-driven receiving zones become syncs and links for more energy-intensive urban load centers. In another scenario, these same areas become populated by socioeconomically exclusive and wealthy enclaves that disproportionately capture the direct benefits of lower long-term energy costs.

These scenarios highlight that if climate destinations, receiving zones, and the U.S. government get this wrong, they face a legacy of exclusion defined by housing and energy precarity. If they abdicate to the momentum of market-driven adaptation, they will be subservient to the whims of market dynamics that may very well advance green growth for some while also driving economic isolation for others. If they get it right, it becomes a model for deep decarbonization and a beacon of optimism in a dark time.

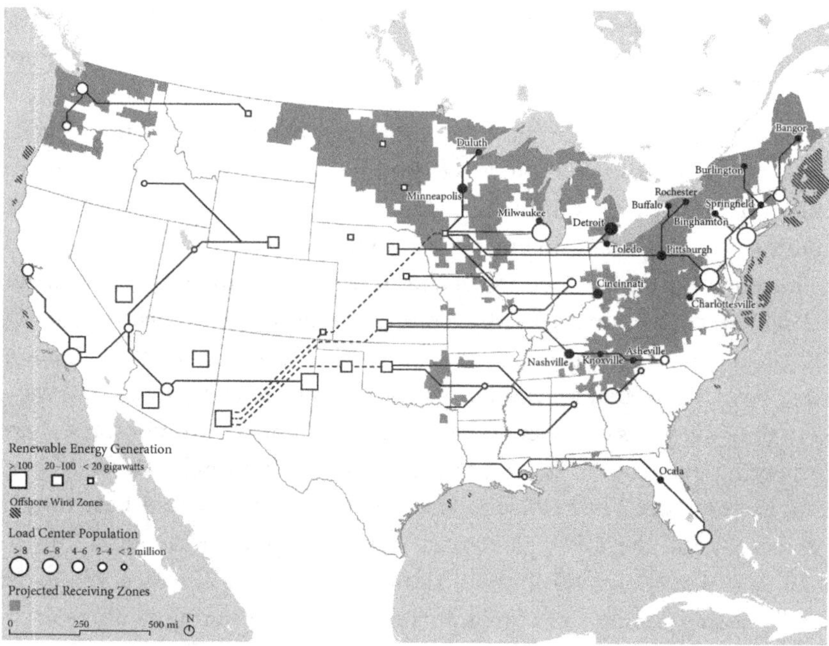

Map 6.2 Potential Convergence of Interregional Renewable Energy Zones and Receiving Zones. Design Credit: Oliver Oglesby and Jesse M. Keenan. Data Source: Hurlbut, David J., Jianyu Gu, Srihari Sundar, An Pham, Barbara O'Neill, Heather Buchanan, Donna Heimiller, Mark Weimar, and Kyle Wilson. 2024. "Interregional Renewable Energy Zones." NREL/TP-6A20-88228. Golden, CO: National Renewable Energy Laboratory; PLACE Initiative. "Receiver Places." https://placeinitiative.org/receiver-places/.

Sketching a Post-Climate America

The future of post-climate America is defined by a point in time when societies and economies have internalized the uncertainty, the risks, and the opportunities of climate change. It is a place where sending and receiving zones are defined with a resolution that affirmatively shapes peoples' aspirations and preferences for where and how they want to live. It is a moment when household capabilities and institutional capacities are reinforced by adaptation policies and plans at every level of government. It is a state where the adaptation of markets is tempered by the goals and values of democratic institutions. It is a time when the decarbonization of America is an unstated assumption in a broader shift in social norms that recognizes both personal and institutional accountability and the plight of climigrants and the immobile alike. In this post-climate America, climigration is not a stigma but a representation of the endurance and continuity of a country built on cycles of immigration and migration. It is the fulfillment of the promise of climate action under a renewed optimism of a shared future built on a balance of American civic nationalism and local autonomy.

This book has highlighted the extent to which the environment, climate, and climate change interact with economic, cultural, and affective considerations to shape the push and pull of migration. In this context, there is emerging evidence that the draw of warm weather and cheap housing is waning in the final years of the era of Sun Belt migration. To the contrary, there is emerging evidence to suggest that a reversal of Sun Belt migration may be on the horizon. The motivations to move to warm weather have largely evaporated, and there is a greater willingness to pay for the environmental amenities of cooler, less humid, and more moderate climates. While the total projected number of climigrants and those relocating is unresolved, the converging evidence and expert opinion are that it will be in the tens of millions of Americans. Even small numbers of climigrants—in the hundreds or thousands—moving to places like Duluth will have an outsized effect on local economies and infrastructure.

Much of this book has focused on the push factors of physical climate impacts and economic stresses. The physical impacts of climate change are vast, from extreme heat and precipitation to the prospects of endemic tropical diseases. Beyond the impacts to one's home and health, the secondary implications for the quality and reliability of public and utility services highlights the prospects of setting back generations of progress in human welfare. It is not just about turning on the lights, as public services for everything from the education of children (without air conditioning) to the provision of public health face being compromised. From increased taxes and utility bills to rolling blackouts and outages, the effects of a declining quality of life are already impacting millions of Americans.

This book has highlighted the complexities of a broad field of adaptation science that offers powerful concepts for describing and understanding how species, people, organizations, institutions, and engineered systems prepare for and/or respond to changes in the environment. From transformation to the many different forms of resilience, each concept offers its own point of view that is critically important for framing the trade-offs of any intervention advanced in the name of adaptation. As has been highlighted throughout this book, adaptation to some may be maladaptation to others.

Adaptation policy and planning in the public sector is well underway in the United States. Even without federal leadership and support under the second Trump administration, state and local agencies are actively engaged in planning for the impacts of climate change. Even under Democratic presidential administrations, the federal government failed to develop a national adaptation strategy or to provide clear signals about the nature of cross-jurisdictional coordination and cooperation that is going to be necessary to drive adaptation investments, much less the management of climigration and relocation. In this context, local governments have been left to their own devices. While many local governments have advanced vulnerability studies and even adaptation plans, a vast majority of jurisdictions throughout the United States lack the institutional capacity, financial resources, and sometimes even the political will to plan for adaptation. This is particularly true for emerging sending zones in the Sun Belt.

By contrast, this book has outlined the emerging evidence that the private sector is driving its own self-interests through rapid advances in intelligence, pricing, and management that represent broader adaptations of the market economy. It is not just firms and portfolio managers that are driving adaptation. Everyday buyers and sellers of housing are increasingly discounting not only the current determinations of physical risk but also the long-term future value of their investments. Climate signals in everything from the muni bond market to the insurance market are getting louder, and this book highlights the political and economic disincentives that sometimes lead politicians and governments to mute these signals.

As these signals get louder, they are already changing the belief systems of local consumers, communities, and even loan officers. Collectively, the changing perceptions and values of market agents are accelerating the adaptation behaviors of markets. The result for consumers in high-risk areas is the prospects of widespread increased costs for housing, mortgages, home and car insurance, and other financial services. Collectively, these increasing costs may very well serve as a powerful push factor either shaping or driving future climigration and relocation decisions. As these processes accelerate with better science, technology, and market intelligence, the chorus of market signals is drowning out the weak mutters of public-sector adaptation. In this regard, private-sector adaptation is driving many of the pathways of future adaptation compromising the capacity of democratic institutions to determine what and who gets saved and what and who gets left behind.

Closer to home, this book has examined how phenomena defining sending and receiving zones are shaping the push and pull of climigration. Sending zones are quickly emerging as part of a broader set of multiscalar processes, which speak to the unwinding of places and communities. This book has highlighted the diversity of behaviors and policies that may shape how local governments either manage relocation or gradually unwind, to their detriment. To compound the challenge for local governments, processes such as climate gentrification and bluelining have been demonstrated to accelerate this broader destabilization of communities and tax bases.

Beyond the economic geography of risk, a variety of actors are coming into focus along a continuum from the episodically displaced to the electively mobile. Even opportunistic investors are part of a broader range of actors that define a shifting landscape for amenity preferences and relocation decision making. Across these different actors, everything from the psychological dissociation of individuals to the spiraling decline of public services is likely to shape the parameters of emerging sending zones.

By contrast, receiving zones and climate destinations are coming into focus in both popular discourse and by virtue of the lived experience of people who are already on the move. Although the popular conceptualization of "climate havens" is potentially problematic, these climate destinations offer a guiding beacon for local adaptation and climate mitigation investments and a positive mental model for future climigrants who seek a better life somewhere else. As this book has highlighted throughout, adaptation is about trade-offs, and there are winners and losers along the way. Climate destinations and receiving zones are endowed with the potential for tremendous opportunities for advancing sustainability in the built environment, but they also bear the burden of providing housing, infrastructure, and public services that are socially and economically inclusive of a wide range of stakeholders. A failure to get this right may result in unsustainable growth patterns that represent the worst of American consumption and social isolation.

The next and final chapter is a work of *fiction* that is informed by the observations and arguments outlined in this book. It gives a face and a name to one of many scenarios that define a post-climate future in America. Along the way, America faces both the benefits and burdens of transformative adaptation. This book is a warning, but it is also a projection of the optimism that is necessary for building and preserving the places and values that define a post-climate America. The next chapter will be written by Americans with the benefit of knowing that there is hope.

7
America 2079

Walter had a secret—he knew what gasoline smelled like.

For as long as he could remember, people plugged their cars in. A few months back, Walter had found some videos online of groups of mostly older men who had figured out how to refine gasoline in their backyards. It seemed like an exceedingly dangerous misadventure just to fire up some old, rusty engines. Yet the spectacle drew him in for reasons that he could not explain.

He wondered why these people had taken these kinds of risks. There was something fundamentally primal at stake in their preservation of this old technology. There also was something missing in Walter's new life upstate. It was too safe. A few hundred dollars later, a flask of bootleg gasoline gave him a sense of belonging.

What these gasoline rebels were risking was not only jail time for the possession of a controlled fossil fuel; they were also risking a major hit to their Climate Credit Score (CCS) that was on par with a personal bankruptcy. Walter knew firsthand how risky behavior could impact his finances. He had recently bought some red meat, and his CCS came down a full 20 points.

"Walter, please. How many times have I asked you to stop watching those vile videos?" asked his wife Mary. "For all you know, Equifax and the Bureau are tracking you. Do you seriously not want to move into the enclave? I'd love for our kids to have access to fresh water through at least high school," she said sarcastically. "Maybe having access to unlimited air conditioning will improve their test scores? I'd really like for them to move out one day."

Their kids had struggled with the move upstate. But they loved being able to go outside, even in the summer. Divorce rates were a good bit lower upstate compared to places in the Sun Belt. Being trapped inside for weeks at a time from May to October often took a toll on families.

Walter and Mary had met in college at Tulane University just before it moved from New Orleans to the rolling hills of Vernon Township, New Jersey. Walter was in the first class of plumbing majors, and Mary had majored in religion and philosophy. They could not have been any different. Mary

North. Jesse M. Keenan, Oxford University Press. © Oxford University Press (2025).
DOI: 10.1093/9780197641644.003.0008

had grown up in Buffalo, and she came from a wealthy family who had made a vast fortune in real estate. "They are all profiteers," she once told Walter. While she rejected her association with the landed gentry, she often spoke with a slight Canadian accent to subtly signal her patrician roots.

Walter grew up in Florida, something he did not tell many people. When immigration was just immigration, America had plenty of skilled workers in the trades. With international climigration and the permanent closing of the southern border in 2040s, plumbers, electricians and contractors of all types were in great demand. Majoring in plumbing science seemed like the quickest way to pay back his student loans and send money back home.

In college, they shared a romantic sense of adventure living on the island of Orleans Parish. The wispy moss in the oaks and the rhythm of the Caribbean expats who had moved into the dilapidated Uptown mansions of sugar barons and oil executives set the scene for their youthful adventures. All-white estates had subdivided and transformed into the bright oranges and greens of the Caribbean. Goats, chickens, and other gator morsels were a frequent sight on the abandoned golf greens of Audubon Park. The holy trinity still fed the masses. There was an endless sense of adventure that only one's youth can tame. In the end, the dengue was exhausting, and they had to get serious about their lives after graduation.

"Honey, what should I wear to our interview with the enclave?" Walter asked while breaking the silence of daily meal preparation. Precisely weighing fresh vegetables and grains every day took more time than they cared to spend together in their narrow kitchen.

"I already put a note on your calendar. This is a big deal. You only get one shot at getting into a good enclave," Mary pleaded. "I don't want you looking down on someone you *think* is a nativist. We need to make a good impression."

"I can fit in, Mary."

"I feel like we are one misstep away from those climigrant people that we see in the park. Can you even imagine what their climate scores might be? Floridians are so sad."

"Hey . . . I'm a Floridian. Ocala—born and bred. Music, sunshine, and good times," Walter argued with the refrain of a lost composition.

"More like lie, deny, and die," snapped Mary. "It is almost like they didn't see all this coming. But we still love you."

Climate Credit Scores

The CCS system was created by several acts of Congress in the final years of the Graham administration in the 2040s. Prior to that, carbon taxes and cross-border trade adjustments were seemingly effective in bringing down the carbon footprint of the country. What few politicians had anticipated is that the dividends that came with the carbon taxes were more or less spent on buying more stuff, and the overall increase in consumption would set back the decarbonization agenda for many years. After international pressure, Congress created a system of social credit modeled after successful systems in Scandinavia and China.

Everything in the supply chain of the economy was assigned a carbon equivalency value. Everyone from producers to retailers reported detailed manifests of carbon equivalency flowing through the economy. At the bottom of the system, trillions of transactions each year were encrypted with keys that were uniquely assigned to each person's tax identification number (TIN). It was the only useful technology to come out of the global shutdown of crypto a decade earlier. To ensure compliance, Congress created the Bureau of Personal Climate Accountability, or the "Bureau" as it was known. Every year, it took tens of thousands of part-time Bureau agents and auditors to keep the system running.

Within just a few years of the CCS going live, the real-time assessment of the carbon intensiveness of goods and services was widely diffused in the economy. Different people paid different prices—depending on who they were, where they lived, and what their climate score was. AI tracking of each TIN helped facilitate this. Consumers could waive statutory privacy protections and allow their movements to be tracked by the Bureau. These waivers led to a one-time bump in one's climate score, but the long-term implications for total compliance were unsettling to many people.

Everything that people bought and consumed weighed on their score. Bypassing the system with cash was minimized with penalties for cash withdrawals. Daily, weekly, and monthly patterns were always under scrutiny. Taking the train to work every month might slightly reduce your score, while buying a regional plane ticket would set a person back a year's worth of good credit. If one were to rather gratuitously buy triple-ply toilet paper instead of single ply, there would be a small but measurable impact on one's score. As Walter was so painfully reminded, lab-grown meat was OK, but actual red meat was a luxury that only few people could justify. McDonald's had long transitioned to meat substitutes, and people gradually lost their taste for charred gristle. Steak simply became the shame-inducing ortolan of the twenty-first century.

The construction of the CCS system was based on a secret methodology determined by economists and social scientists at the Bureau. It was commonly just referred to in the press as the "Taxonomy." The Bureau's management of the Taxonomy had always been viewed as a work in progress. Senator Daniel Ismay's infamous remarks on his intention to "break" consumers resonated as both political theater and a matter of fact. Congress was sensitive to the proposition that the CCS would never work if it moved too fast. By law, each quintile of the population must fall into one of five classifications as a means to benchmark consumption. Over time, it would become harder every year to move your way into the top quintile of lowest impact because the score was adjusted to increasingly penalize carbon-intensive consumption. The Taxonomy could not be too draconian, or it would entirely cut off the consumer economy. This often meant that the Taxonomy was adjusted, and benchmarks were reset in response to cycles in the real economy and the Federal Reserve's agenda.

When ideas of social justice were still fashionable in the United States, the Bureau sought early on to address existing socioeconomic inequities through the standardization of scores that reflected uneven starting points among consumers. In this sense, someone living in rural America without access to mass transit would not be penalized for gas purchases or, later, electric vehicle (EV) charges. Fossil fuels would finally be phased out and ultimately outlawed a decade after the CCS system was implemented. Initially, each rural American consumer was allocated an equivalent adjusted gas mileage credit standardized for a fixed allocation of vehicle miles traveled (VMT) per year. While credit exchanges became popular for people who did not travel their allocated VMTs every year, the Bureau eventually instituted personal biometric validators at every gas station and EV charging station in the country. The bootleg energy market proliferated for a while, but it merely operated to drive the demand for personal energy storage devices that were exempted from CCS reporting.

An additional challenge arose for historically marginalized and low-income communities who found that climate scores were merely another cost burden, particularly for those households who did not have the resources to live carbon-neutral lifestyles. The Bureau attempted to remedy this problem by allocating non-transferable climate credit points for a wide variety of households. For instance, individuals living in environmental justice (EJ) and tropical endemic disease (TED) zones received credit bonuses that more than offset any additional imputed carbon tax. Through a later amendment to the original legislation, individuals over the age of 65 living in EJ and TED zones could redeem some portion of their climate credit points for a premium reduction on public healthcare plans.

The healthcare economy had always been a challenging sector for the CCS Taxonomy because keeping people alive requires a tremendous amount of resources. The original legislation had taken steps to exempt nearly the entire healthcare economy from CCS reporting. This was a massive blow to carbon mitigation efforts for a number of reasons. One major problem was that, with a rapidly aging society, many people claimed their legal residence as an aging-in-place healthcare facility. While this loophole allowed the aging population's consumption to be largely exempt from the CCS system, it greatly reduced the amount of available housing on the market. A major consequence of this overconsumption of housing was that most senior citizens—the largest segment of the population—had a much higher carbon footprint than younger generations.

Unfortunately, this was the least of the housing market's concerns. The Great Climate Mortgage Crash of 2032 had set the stage for a highly dysfunctional housing market that the CCS Taxonomy would try to correct. In 2032, the Federal Housing Finance Agency (FHFA) implemented a new untested technology that evaluated, with a great deal of imprecision, the physical climate risks impacting any given property within the term of a 30-year mortgage. Upon implementation of the program, global investors in Fannie Mae and Freddie Mac securities—aided by the credit rating agencies—imposed a strict set of risk premiums.

This led to a steep overcorrection in mortgage pricing that dried up an estimated 30% to 40% of available mortgage capital in most coastal and wildfire-sensitive housing markets. The resulting collateral devaluation led to a foreclosure crisis that accelerated the spiraling devaluation. Congress was delayed in stabilizing the market because of post-Trump austerity measures stemming from the 2025 tariff debacle, as well as the poor optics of bailing out rich coastal property owners. By the time the Federal Reserve's asset purchases stepped in to stabilize the market, it was too late—the Minsky moment had come and gone.

With the Great Climate Mortgage Crash of 2032, there was a lot less available mortgage financing to go around. One consequence was that the transfer of housing within families became much more important. Unfortunately, this had the effect of limiting economic and geographic mobility, as people were constrained in their ability to purchase housing on the open market because of severe supply constraints. This immobility had severe consequences for the labor market, as it coincided with international climigration being shut down. As economic and labor mobility declined, regional imbalances in labor and housing markets made it increasingly difficult for domestic climigrants to relocate. One immediate result of this labor imbalance was that low-wage industries, such as meat-processing plants, crowded into high-risk coastal geographies where people were desperate for work.

In spite of these economic stresses, the housing industry was a major target of the Bureau. Any single housing unit sitting on greater than a quarter of an acre (.10 hectare) of real property was heavily penalized by the CCS Taxonomy. To accelerate the decline of wasteful lawns, the purchase of turf management products would be penalized. Almost overnight, green grass began to disappear from the landscape. In addition, most people could not afford more than 50 gallons (189 liters) of water per capita a day, given the significant increases in water utility rates. As sustainability measures, carbon taxes, and CCS-driven behavior significantly reduced water demand, utilities exponentially raised rates to cover lost revenue and their long-term debt service and capital investment obligations in the infrastructure. Gray water systems for irrigation were popular for a time, but many public health boards began to regulate cisterns after a series of serious water-borne disease outbreaks prompted a crackdown.

In the 2020s, the extraction of sod and turf was the one of the largest export markets by weight in the United States. By 2039, that export market had entirely disappeared. The sod and turf producers simply turned to soil banking and sold carbon tax credits. The industry was so profitable that leeway was given to exempt much of their activity from various carbon taxes. Unfortunately, once soil theft started taking place, it became extremely difficult to get a federal soil bank license. While the soil carbon industry was booming, it meant that American landscapes were often devoid of decorative plants. Composting bins and vegetable gardens replaced many ornamental gardens, even in public spaces. By the 2040s, public parks were places of both leisure and public food production. For this reason, parks often attracted destitute climigrants who were more than willing to work for their food.

The physical and metaphorical landscape of cities dramatically changed, in part, because of the CCS Taxonomy. These transformations were facilitated by a parallel change in federal income tax policy that removed many of the long-term capital gains benefits of owning and developing real estate. The tax policy changes, together with the Great Climate Mortgage Crash of 2032, removed a lot of leverage and corresponding speculation in the housing market. Much of that capital migrated over to sustainable building materials and construction systems that benefited from modified accelerated depreciation schedules and passive loss allowances. Essentially, the government allowed passive losses for sustainable housing investors to stimulate the growth of the market. For a few years, this drove up the production cost of housing, but bulk purchasing by contractor co-ops and enclave developers ultimately stabilized prices. The cost of capital for sustainable housing in exclusive enclaves was never more than a narrow spread over 10-year U.S. Treasuries. Although money for sustainable housing was cheap, land in the Northeast and Midwest was scarce, and the overall supply of housing struggled to keep up with population growth.

Between carbon taxes and CCS penalties, almost all new housing production was carbon neutral in terms of operational energy by the 2050s. The larger challenge was narrowing down carbon within the embodied energy of housing and buildings. The vast amount of debris from the coastal and Western housing stocks made decarbonization challenging, but it also accelerated the production of higher-quality housing. To simplify the CCS implications for buying and renting, the Bureau developed a consumer-facing online platform that allowed prospective property owners or sellers to have their property graded from A to F. The grading scheme utilized large language models built on government databases to estimate a variety of factors, including material and transportation emissions. Unfortunately, many lower-income people could only afford to rent Grade D and F properties. The higher utility costs together with the negative CCS implications often worked to reinforce cyclical poverty.

Size mattered. Per capita indoor space allocations exceeding 450 ft^2 ($42m^2$) were heavily penalized in the CCS Taxonomy. This had the unintended consequence of slightly reducing birth rates in the suburbs for a number of years. Birth rates somewhat rebounded once housing prices began to reflect the economization and climate credit benefits of greater densities. Hardly anyone could afford to live in an oversized home, especially near the coast. Not only did prices for the remaining stock of oversized homes drop, but much of the residual market value was based on the land and the scrap value of the materials in the house. Insurance, taxes, and utilities for these oversized homes were often many times the monthly mortgage payments, if prospective buyers could get a mortgage at all.

Properties with large lots were redeveloped and divided into multiple condominium units. For properties with extra-large lots, the developer could either dedicate a certain portion of the lot as privately owned public spaces (POPS), or they could develop more affordable housing units. Some developers choose to build POPs with real grass as a luxury amenity to attract wealthy buyers.

This redevelopment process gave rise to "enclaves" of dense planned unit developments that were built on the foundations of previously inhabited exclusive golf and country club communities. To combat the possibility of unlimited intra-family transfers, most of these developments were run by restricted co-ops who curated a precise composition of corporate shareholders—or neighbors. Even simple condominium developments offered only life estates instead of fee simple titles to enforce an emerging social caste system within the enclaves. This often meant that people owned the homes only for their or their children's natural lives. A life estate would only be granted to a subsequent generation that was compliant in all aspects of the social and environment order. While these enclaves were packed with cutting-edge sustainable technologies and infrastructure, they often mimicked the strict social hierarchies of the communities that they replaced.

The Interview

"Thank you so much for having us, Mr. Lambert," said Mary, who feigned an awkward curtsy. Walter was too distracted by the eco-modernist and new urbanist architecture to notice.

"Well, we are thrilled to have you all. Building a new world, much less a new community, requires like-minded people sharing our common philosophy for sustainable action," said Mr. Lambert, who gestured to an all-white building nestled in the wooded portion of the enclave. "Our town hall was modeled after the town hall in the new urbanist community of Seaside, Florida, that was unfortunately lost to the sea some years ago. But, I promise, that is the only reference to Florida that you will have to hear about today."

"May I offer you all some recycled water? We brew it right here at Rosemary Peak."

Throughout the course of the morning, Mary and Walter were joined by other people for a tour of the environmental technologies and amenities at the enclave. They did not anticipate other people being on the tour, and they somehow felt belittled by the inclusion of others. They were not competing against themselves. They were competing against others.

All the other couples were wearing complementary earthen hues. Mary and Walter felt over-dressed and unmatched. At every stop, the four couples and one graying woman from upstate were prepared with technical questions about energy, water, waste streams, and food production. They even referenced their questions in the metric system. "Who are these people?" Walter silently mouthed.

At lunch, Mary and Walter were assigned a table with a couple from Atlanta. "So, what brings you folks upstate?" asked Walter with the best semblance of a gentile Florida accent that he could muster. "Don't be so nosy, honey. Everyone needs drinking water," Mary interjected.

"Actually, we left Atlanta before they ran out of water. It just wasn't a place where we wanted to raise a child. The climigrants really changed the culture, and don't get me started on the traffic," said the man. "You know some of those climigrants came from countries where they drive horse-drawn carts rather than EVs, and their lack of a sense of spatial depth causes all kinds of accidents. Of course, people from this country aren't any better. I don't think anyone in Louisiana every used a turn signal."

"Darling, what did I tell you? We have to change our perspective on life and the people around us. At Rosemary Peak, we are all in this together," said the man's wife.

"I just hope that people in Rosemary don't throw their trash in the street and play their music at all hours of the day. I wouldn't have any problems with climigrants if they just obeyed the rules," said the man attempting to defend himself.

"David Franklin Jones, darling, we *are* climigrants. The faster that you come to terms with it, the faster you will find some happiness in your life—our life, our family's life."

The cathartic frankness of David's wife fully captured Walter's attention. He wanted liberation, too. Mary was quick to change the subject, knowing the man's bigoted outburst would disqualify them from the competition.

After lunch, Mary and Walter found themselves in Mr. Lambert's office. It was a home office with a Murphy desk in the living room that opened onto the patio to accommodate the space needed for meetings. The living room was decorated with maps of Martha's Vineyard, nautical charts, and framed pieces of scrimshaw. One frame over the electric fireplace contained a segment of a wood exterior panel from an old building. "Is that from your hometown?" asked Walter.

"Yes, thank you for asking. That is the last piece of construction from the Old Whaling Church in Edgartown known to exist. I bought it at a charity auction that supports the final relocation of the last inhabitants of the island by Community Services. It reminds me of home."

"So sad," lamented Mary.

"No. It is not sad. They were warned. The billionaires lost their estates, but the Wampanoag lost their ancestry. That is the real crime. It's not sad, it is infuriating. My family were there for only a handful of generations, but we have nothing to claim from that place other than our memories. The Wampanoag had something sacred stolen from them."

"Listen, I've been wanting to talk to you both," said Mr. Lambert. "Let me just say this—we here at Rosemary Peak are excited to have you join our community. We feel that your disciplined way of life, public service, and background have prepared you for this opportunity."

"What do you mean *background*?" asked Walter.

"Well, I've known Mary's father for decades. Real estate is an awfully small world. But I will say, we were surprised to see that you are from Florida, Walter. You don't come across as a Floridian. Were you born there? I didn't see your birth certificate and apostille in your application file."

The Great Florida Partition

Walter grew up in Ocala, Florida. After college, a series of events would transpire that would prevent him from ever wanting to return home. On October 20, 2057, Hurricane Grant made a direct landfall just south of Hurricane Pass on Clearwater Beach. A Category 6 hurricane had never made a direct hit before on the Tampa Bay area. Category 6 hurricanes had only been classified as mega-hurricanes by the World Meteorological Organization a few decades earlier. With 220 mph (98 mps) sustained winds at its core, the right lower quadrant of Grant drove almost 59 ft (18 m) of storm surge into Tampa Bay. More than 10,000 people drowned, and another 4,100 bodies were never recovered. The loss of life would have been greater, but the region had been slowly moving to higher ground with sea level rise (SLR). Waterfront enclaves in places like Temple Terrace were able to weather the storm, as these upscale enclaves were designed to withstand a direct hit and operate autonomously for up to a year. Many other communities were lost to a morass of mangroves and housing debris that marked the site of a once great metropolis. For years, those who survived would swear that they saw ghosts and could even smell human decay in the dark red mangroves that had migrated north into the ever-rising shoreline.

Over the prior decades, the triangle mega-region of Orlando-Tampa-Gainesville (OTG) had become the primary destination for climigrants from South Florida, as well as legal and undocumented climigrants from the Caribbean. By 2056, 80% of Florida's GDP and 70% of its population were located in the OTG mega-region. Climigration had been good for business. In the 2040s, 400 acres (162 hectares) were bulldozed every day for housing development. The CCS Taxonomy later made housing development in much of the OTG too expensive, given the risks of SLR and hurricanes. As a workaround, the State of Florida created a state-owned housing finance agency and issued mortgages for tens of billions of dollars to fuel new housing development.

Thanks to a positive phase of Arctic Oscillation and a weakened jet stream that would have otherwise quickly pushed a hurricane out into the Atlantic Ocean, Hurricane Grant sat over the OTG for a week blowing off roofs and dropping close to 3 ft (1 m) of rain. The devastation was vast. Within a month, 5 million people had left Florida. Some experts noted that the population influx into Atlanta is what ultimately doomed the region's drinking water system. For the millions who remained, the mix of experiences varied from those living rather comfortably in enclaves to those that lived in one of dozens of camps. Children would be born and go through high school without moving out of a Grant camp.

Hurricane Grant would set in motion a series of events that few could have foreseen. First, the Florida insurer of last resort—Citizens Property Insurance Corporation—would become insolvent and attempt to file Chapter 9 bankruptcy. So too did the Florida housing finance agency. A bankruptcy of this type was rejected because the appellate courts and creditors viewed both entities as units of the State of Florida and not municipalities eligible for bankruptcy protection. States cannot lawfully file for bankruptcy. The resulting litigation tanked the state's credit rating. Upon final appeal to the U.S. Supreme Court, the court ruled that the State of Florida was functionally insolvent and would be forced to liquidate assets to pay the tens of billions that it owned insured parties and its creditors from Hurricane Grant losses.

Because the state was rated with a junk bond status, it was severely limited in how much money it could borrow. As it turned out, legislators in Tallahassee had not purchased nearly enough reinsurance to cover Citizens. Instead, they pocketed the occasional surplus to spend on pet projects and enclave infrastructure enhancements for their wealthiest constituents. To make matters worse, they were also sold bogus unregulated derivative products that did not pay out a single dollar in the end. With bills to pay and without the capacity to borrow money in the near term, the state had few options.

As a result, the state severely limited public services and proceeded to privatize a great deal of the state's assets and infrastructure. Tallahassee even sold the University of Florida to the DeVry Corporation, the nation's largest and most well respected AI-instructed university system. Once ranked one of the top public universities in the country with live instruction, the University of Florida had become a listless flagship, with successive swings between conservative and progressive governors tinkering with the curriculum and direction of the institution. Everything was sold to the highest bidder, even if the highest bidder was an affiliate company owned by members of the Florida legislature. Water districts, fleets of Florida Fish and Wildlife Conservation Commission boats, and even some of the state forests on the high ground of the Panhandle were sold off. The political backlash was severe, and Republicans were voted out of every statewide office within a few years. The progressive Democrats who took over only made problems worse.

In times of trouble, Florida had historically turned to its tourism economy for additional revenue. Unfortunately, tourism in Florida had been on a long decline. There were very few sandy beaches left to handle what was left of the nearly 200 million tourists that Florida used to accommodate every year. With SLR and a shortage of sand, only a handful of recreational beaches remained in Florida. Elevated concrete pads with microplastic sand were popular for a while, but these beachside structures eventually just became barren concrete landscapes. People increasingly fell out of love with going to the beach because it was just too hot to go outside between 9 a.m. and 8 p.m. for half the year. In the shallow Gulf of Mexico, water temperatures were often over 100°F (38°C), and this presented a serious health hazard in the summer. Pools did not offer much comfort either. While some enclaves used cool condensate from their district-scale heat pump systems for pool water, most peoples' climate credit score would not allow them to "ice" their pool. On special occasions, such as July 4th, people would throw a few dozen bags of ice into their pools for the ultimate relief.

Sport fishing had also largely disappeared for a variety of reasons, including the fact that many fish were simply being boiled alive in the warm near-shore waters and were forced to swim to deeper depths farther out at sea in less nutrient-intensive areas. The last known hogfish—a local delicacy—was spotted in the Keys decades ago. Long gone were the days of fishing off the pier for snapper, or any fish for that matter. As a result, almost all local restaurants sold frozen seafood from Southeast Asian fish farms or illegal fisheries off the coast of East Africa. The era of family-friendly seafood in Florida was over.

The theme parks had not fared much better, particularly with increasing media attention on heat-related deaths of tourists. Intravenous fluid retail stores popular since the 2020s were no longer just for college kids. After successive waves of dengue and zika VII, tourists simply did not want to put their children at risk. To make matters more challenging, tourists often found it hard to find a hotel room. Many of Orlando's 150,000 hotel rooms were occupied by people displaced by Hurricane Grant. Eventually, the Comcast Continent replaced Disney World without much fanfare or ambition. Disney's new park on the shores of Lake Champlain in Vermont would become the magical destination that kids would grow up dreaming about. The silver maple (*Acer saccharinum*) would replace the royal palm (*Roystonea regia*) in the zeitgeist of American leisure.

Desperate for tax revenue, the progressives in Tallahassee reacted to Orlando's theme park decline by turning Orlando into a city of vice somewhere between Las Vegas and the long-vanished Atlantic City. Prostitution, violent crime, and illegal human trafficking of climigrants followed. Long cigar-shaped speedboats similar to those that once trafficked cocaine from the Bahamas in the 1980s were now filled with undocumented foreign climigrants on midnight runs from one of the Bahamas' last two remaining fortified islands to the coast of Orlando near Port Alafaya. In less than a decade, Orlando would go from a family-friendly tourist destination to a hollowed-out shell of massage parlors and musty casinos. The smell of sweat and the faded turquoise were clear indicators of where Orlando had been and where it was destined to go.

Gambling and hotel occupancy taxes could not save Florida's public finances. Even Florida's agricultural economy was struggling. Strawberry farms in southern Dade County closed because of saltwater intrusion. A citrus disease spread north from Brazil that devastated the Central Florida crop yields. Unprecedented severe winter freezes compromised everything from green beans to tomatoes around Marianna. Sugar cane was deemed too carbon intensive and had given way to housing development for climigrants from the south. Like the housing industry, the agricultural economy suffered without foreign migrants with the skills necessary to sustain the fields and crops.

Florida was not just missing workers; it was also missing retirees. Some in Tallahassee called for higher portfolio and estates taxes, but retirees had left Florida a generation before. The final death knell in Florida's retirement economy came when health insurers were allowed to exclude coverage for chronic health conditions from tropical diseases like malaria in TED zones. The resulting cost of full coverage health insurance, together with already unaffordable home and car insurance, was too much for retirees on a fixed income.

Without retirees, tourists, and sugar, Florida was insolvent and running out of money after Hurricane Grant. Even the most basic services like issuing birth and death certificates were expensive and not consistently administered. Climigrants often found it difficult to prove that they were once Florida residents. Without birth certificates, it was hard to prove that they were U.S. citizens. The illegal market for these official documents flourished in this vacuum. Florida began to lose track of how many people actually lived in the state. One thing was clear—the population was rapidly declining by hundreds of thousands each year, and foreign immigration and climigration were no longer driving net increases in population, as they had for nearly the past century.

The one thing that legislators in Tallahassee had feared the most was just on the horizon—a state income tax. In 2062, almost 40% of registered voters called for a constitutional convention. The sole task at the convention was to design an income tax for the state. The convention was held at the Orlando Convention Center. This was not without controversy. While the progressives who represented the OTG's power establishment felt that Orlando was the true center of the state, the Christian nationalists who dominated the politics in the northern part of the state resented the very existence of Orlando. Their media channels had long broadcast warnings about "Sodom and Gorlando," as they called it. As the convention approached, convoys of North

Florida militia in EVs escorted delegates into the city and proceeded to set up secure camps in abandoned parking lots of former theme parks.

The convention was dominated by two opposing forces—north and south. The Southeast Florida coastal counties of Miami-Dade, Broward, and Palm Beach had been consolidated back into Monroe County. Lee County had consolidated much of Southwest Florida. This political bloc of once great coastal counties was allied with the OTG in their singular ambition to impose a state income tax. They desperately needed the tax revenue to maintain their expensive climate adaptation projects and urban infrastructure. However, much of North Florida and the Panhandle is on high ground, and the political and economic momentum of the state was clearly moving north. North Floridians resented having to pay for the regular reconstruction of sinking South Florida cities. They were made to believe that South Floridians were a lawless cohort of foreign climigrants who would carry forward a challenge to their cultural values and economic way of life. Orlando was not just the site of the constitutional convention; it was also on the frontlines of a culture war.

Within a few days of the start of the convention, the delegates had pushed through a new constitutional provision for a state income tax. Within hours of the news, black smoke could be seen on the horizon around the various lakes of Orlando. The distant sounds of automatic gunfire and shock grenades seemed to get closer by the hour. The state police under the command of the progressive governor had attempted to create a green zone around the convention center, but this only fueled the flames of misinformation. While there would later be competing historical accounts of what happened next, the end result was open armed conflict between the North Florida militias and the outgunned state police.

The governor called in the National Guard, but they were not prepared for urban combat, as their missions had largely been relegated in the past generation to post-disaster search and rescue. Many soldiers were deployed without ammunition. The resulting casualties from the state police and National Guard were much greater than the general public could absorb. Eventually, a reluctant president, with the consent of Congress, deployed the 1st U.S. Marine Division under the Posse Comitatus Act to quell the insurrection. By this point, the militia were largely out of ammunition and ready to plead their case in the court of public opinion.

The convention was able to carry on its business throughout this armed insurrection. In its final act, the convention dissolved the Constitution of the State of Florida. In a compromise negotiated by a bipartisan group of U.S. congressional leaders and supported by nearly all the states, the convention was divided between North and South Florida delegates, who each ratified a new constitution for the State of North Florida and the State of South Florida, respectively. Map 7.1 outlines the borders of the two states, with the OTG region serving as the economic and political heart of the State of South Florida. This Great Florida Partition would mark the end of the culture wars that plagued so much of American politics, but it would also mark the beginning of a rapid exodus from both North and South Florida. Florida had long been a fiction of leisure marketing, but when it had nothing left to market, it could no longer sustain its own existence.

Map 7.1 Maps of the State of North Florida and the State of South Florida. Design Credit: Oliver Oglesby and Jesse M. Keenan.

Passing, Liberation, and Gasoline

Life in Rosemary Peak for Walter and his family was everything they thought it would be. Summers inside were not so bad with consistent air conditioning. Even the occasional wildfire smoke inundation did not feel so debilitating with the high-tech MERV filters built into every room. The water tasted good, and the pool was for swimming. The enclave deposited money into their account every month for their share of electricity that Rosemary Peak sold to people living outside the enclave. This allowed them to save for the future and to afford the occasional fruits and vegetables from California. Mary and Walter had known very little stability as adults. They were uncertain how they should feel.

But everything has a price. While the kids made friends, Walter began to question what their kids were picking up from other kids in the enclave.

"Patrick, where did you learn that word 'slim-y-grant'?" Walter asked.

"Dad, I don't know. Some of the kids at school were calling Eduardo a slim-y-grant. We thought it was his name," Patrick slyly responded.

"Well, I don't want you to ever use or repeat that word. How do you think that makes Eduardo feel? I can't go back home. Does that make me a slim-y-grant?"

"No," Patrick said reluctantly while avoiding eye contact with his father.

Walter started to wonder whether Patrick was ashamed of his actions or his own father.

Mary fit right in. Although she had desperately sought to remake herself as a moral crusader atoning for her family's wealth, she spoke the same language as her neighbors. She knew the subtle body language that signaled when a conversation could shift from niceties to substance. She knew never to wear logos or clothes with any pattern that signaled their privileged lives. She knew the order and their place within it.

Walter tried his best to make friends. They played bridge and went to their obligatory Garden Club meetings. The people seemed approachable only to slide into the same nativist bigotry that his kids were picking up at school. Few ever expressed gratefulness or an obligation for public service. Instead, their bounded condition in the enclave was viewed as the byproduct of a failed political system and carbon-intensive climigrants living outside the rules of a corrupt CCS system. They followed the rules, why couldn't others? Walter always seemed to stop himself from pointing out that the enclaves throughout the North and South made the rules and enforced them through their status as unregulated utilities and political donors. They controlled the power literally and figuratively.

Things tended to turn darker on the rare occasion that people consumed alcohol. Living in any enclave meant a puritanical order of monitoring and reporting of personal consumption. This order imposed a kind of modesty that tended to curb those who held their own self-interest above that of the enclave. Those who fell out of order could be publicly shamed or even temporarily banished from the enclave. Several households had lost a family member to banishment—almost always young men. Mary and Walter's son had been close to a young man on "sabbatical." Living in any enclave meant enduring the constant tension of being under observation. When people did drink socially, they got drunk. Because alcohol was a rare luxury, it only took a drink or two for people to let down their guard.

"Walter, what's up with you? You always seem like you are having a bad day. Don't you loooove it here in paradise?" questioned one of the neighbors.

"More like Paradise, California," Walter mumbled under his breath.

"I miss Florida. I miss the freedom," he said with a clearer tone of defiance.

"I want to get lost. I want to swim in the Gulf. I want to BBQ with an *actual* fire."

"You are starting to sound like a climigrant, Walt," chimed another neighbor from across the pool. It took a minute for the group to process the direct nature of the insult, as many had wondered how committed their new neighbor really was to the enclave.

"Honestly, fuck you, Rand. I *am* a climigrant. Better yet—I am an *illegal* climigrant," exclaimed Walter. "I might have grown up in Ocala, but I was born in the British Virgin Islands."

A hush settled over the deck.

"After my mother died of dengue, my father brought us to this country so that we wouldn't starve to death. How long have you ever gone without fresh water? You know what? We were once wealthier than this entire enclave. We used to have two slips at the Bitter End Yacht Club. But mother nature took every God damned dollar that we had. We worked hard to get to where we are today. If it wasn't for plumbers, you wouldn't have any of your precious water to drink. Does anyone here want to service their own plumbing? I didn't fucking think so."

The men stood silent for a long two minutes, as Walter caught his breath. He thought for a second that maybe Mary might have overheard his outburst. She did not.

"Listen, Walt. I have never told anyone this, but I am from Corpus Christi, Texas. I am not really from upstate," said his neighbor in a starkly different accent.

Looking upward and exhaling, another man replied, "My family is not from Rochester. My family is from St. Pete. My wife . . . she is from . . . the Caribbean."

Without saying a word, the three men were bound to a mutual pact of secrecy and a common sense of liberation.

"We are all upstaters now," Walter said. "Let's fire up the grill and make some real burgers. I've got a little bit of gasoline that will make it hot, hot, hot. The charred the better. Just the way we Americans like it."

Notes

Introduction

1. Lafarga Previdi, Irene, Michael Welton, Jazmín Díaz Rivera, Deborah J. Watkins, Zulmarie Díaz, Héctor R. Torres, and Chrystal Galán. 2022. "The impact of natural disasters on maternal health: Hurricanes Irma and María in Puerto Rico." *Children* 9(7): 940. https://doi.org/10.3390/children9070940
2. Keenan, Jesse M. 2021, August 31st. "Post-Ida failures show why we need to adapt." CNN. https://www.cnn.com/2021/08/31/opinions/post-ida-failures-show-need-to-adapt-keenan/index.html

Chapter 1

1. Lustgarten, Abrahm. 2024. *On The Move: The Overheating Earth and the Uprooting of America*. New York: Farrar, Straus and Giroux.
2. Adger, William Neil, Sonja Fransen, Ricardo Safra de Campos, and William C. Clark. 2024. "Scientific frontiers on migration and sustainability." *Proceedings of the National Academy of Sciences* 121(3): e2321325121.
3. Nabong, Emily C., Lauren Hocking, Aaron Opdyke, and Jeffrey P. Walters. 2023. "Decision-making factor interactions influencing climate migration: A systems-based systematic review." *Wiley Interdisciplinary Reviews: Climate Change* 14(4): e828. https://doi.org/10.1002/wcc.828
4. Das Sharma, Asmeeta, Gregory Bracken, and Veran Balz. 2020. "Environmental migration and regional livelihood planning: A livelihood planning approach to circular migration." *Environmental Justice* 13(5): 173–180.
5. Ghosh, Amitav. 2021. "The great uprooting: Migration and displacement in an age of planetary crisis." *The Massachusetts Review* 62(4): 712–733. https://doi.org/10.1353/mar.2021.0158
6. Silke, Andrew, and John Morrison. 2022. "Gathering storm: An introduction to the special issue on climate change and terrorism." *Terrorism and Political Violence* 34(5): 883–893.
7. United Nations High Commissioner for Refugees (UNHCR). 2024. *Focus Area Strategic Plan for Climate Action 2024–2030*. Geneva: UNHCR.
8. United Nations High Commissioner for Refugees (UNHCR). 2024. "Refugee data finder." Geneva: UNHCR. https://www.unhcr.org/refugee-statistics/download/?url=IAr67y
9. Rigaud, Kanta Kumari, Alex De Sherbinin, Bryan Jones, Jonas Bergmann, Viviane Clement, Kayly Ober, Jacob Schewe, Susana Beatriz Adamo, Brent McCusker, Silke Heuser, and Amelia Midgley. 2018. *Groundswell: Preparing for Internal Climate Migration*. Washington, D.C.: World Bank. https://doi.org/10.7916/D8Z33FNS
10. U.S. Customs and Border Protection. 2025. "Southwest land border encounters." https://www.cbp.gov/newsroom/stats/southwest-land-border-encounters
11. U.S. Customs and Border Protection 2025.
12. Singer, Audrey, and Sylvia L. Bryan. 2023. "U.S. Border Patrol encounters at the southwest border: Fact sheet." Washington, D.C.: Congressional Research Service. https://crsreports.congress.gov/product/pdf/R/R47556
13. Refugee Council USA. 2025. *US Refugee Admissions*. https://rcusa.org/resources/fy24-annual-refugee-arrivals-report/
14. Linke, Andrew, Stephanie Leutert, Joshua Busby, Maria Duque, Matthew Shawcroft, and Simon Brewer. 2023. "Dry growing seasons predicted Central American migration to the US from 2012 to 2018." *Scientific Reports* 13(1): 18400. https://doi.org/10.1038/s41598-023-43668-9
15. Nawrotzki, Raphael J., Fernando Riosmena, and Lori M. Hunter. 2013. "Do rainfall deficits predict US-bound migration from rural Mexico? Evidence from the Mexican census." *Population Research and Policy Review* 32: 129–158. https://doi.org/10.1007/s11113-012-9251-8

16. Jessoe, Katrina, Dale T. Manning, and J. Edward Taylor. 2018. "Climate change and labour allocation in rural Mexico: Evidence from annual fluctuations in weather." *The Economic Journal* 128 (608): 230–261. https://doi.org/10.1111/ecoj.12448
17. De Sherbinin, Alex, Susana Adamo, Ama Francis, Bryan Jones, and Jane Mills. 2021. *Climate Change and Its Impact on Urbanization in Mexico and Central America.* Report prepared for the Mayors Migration Council. New York: Center for International Earth Science Information Network (CIESIN), the Earth Institute, Columbia University.
18. White House. 2021. *Report on the Impact of Climate Change on Migration.* Washington, D.C.: Executive Office of the President.
19. White House 2021.
20. 8 U.S.C. § 1101(a)(42).
21. Atapattu, Sumudu. 2020. "Climate change and displacement: Protecting 'climate refugees' within a framework of justice and human rights." *Journal of Human Rights and the Environment* 11(1): 86–113.
22. Berchin, Issa Ibrahim, Isabela Blasi Valduga, Jéssica Garcia, and José Baltazar Salgueirinho Osório de Andrade. 2017. "Climate change and forced migrations: An effort towards recognizing climate refugees." *Geoforum* 84 (August): 147–150. https://doi.org/10.1016/j.geoforum.2017.06.022
23. Brzoska, Michael, and Christiane Fröhlich. 2016. "Climate change, migration and violent conflict: Vulnerabilities, pathways and adaptation strategies." *Migration and Development* 5(2): 190–210. https://doi.org/10.1080/21632324.2015.1022973
24. Kamel, Yumna. 2023, November 3rd. "Climate-induced displacement and legal protections." Presentation at Climate Displacement Workship, Boston University, Boston, MA.
25. Bronen, Robin. 2009. "Forced migration of Alaskan indigenous communities due to climate change: creating a human rights response." In *Linking Environmental Change, Migration & Social Vulnerability*, edited by Oliver-Smith, Anthony, and Xiaomeng Shen, 68–73. Bonn: United Nations University, Institute for Environment and Human Security. https://collections.unu.edu/eserv/UNU:1879/pdf4019.pdf#page=70
26. Mallick, Bishawjit, Chup Priovashini, and Jochen Schanze. 2023. "'I can migrate, but why should I?'—voluntary non-migration despite creeping environmental risks." *Humanities and Social Sciences Communications* 10(1): 34. https://doi.org/10.1057/s41599-023-01516-1
27. McLeman, Robert. 2018. "Thresholds in climate migration." *Population and Environment* 39: 319–338. https://doi.org/10.1007/s11111-017-0290-2
28. Hauer, Mathew E., Sunshine A. Jacobs, and Scott A. Kulp. 2024. "Climate migration amplifies demographic change and population aging." *Proceedings of the National Academy of Sciences* 121(3): e2206192119. https://doi.org/10.1073/pnas.2206192119
29. Hauer et al. 2024.
30. Beyer, Robert M., Jacob Schewe, and Guy J. Abel. 2023. "Modeling climate migration: Dead ends and new avenues." *Frontiers in Climate* 5: 1212649.
31. Schewel, Kerilyn, Sarah Dickerson, B. Madson, and Gabriela Nagle Alverio. 2024. "How well can we predict climate migration? A review of forecasting models." *Frontiers in Climate* 5: 1189125. https://doi.org/10.3389/fclim.2023.1189125
32. Hauer, Mathew E., Jason M. Evans, and Deepak R. Mishra. 2016. "Millions projected to be at risk from sea-level rise in the continental United States." *Nature Climate Change* 6(7): 691–695. https://doi.org/10.1038/nclimate2961
33. Klabunde, Anna, and Frans Willekens. 2016. "Decision-making in agent-based models of migration: State of the art and challenges." *European Journal of Population* 32: 73–97. https://doi.org/10.1007/s10680-015-9362-0
34. Beyer, Robert M., Jacob Schewe, and Hermann Lotze-Campen. 2022. "Gravity models do not explain, and cannot predict, international migration dynamics." *Humanities and Social Sciences Communications* 9(1): 56. https://doi.org/10.1057/s41599-022-01067-x
35. Piguet, Etienne. 2010. "Linking climate change, environmental degradation, and migration: A methodological overview." *Wiley Interdisciplinary Reviews: Climate Change* 1(4): 517–524. https://doi.org/10.1002/wcc.54
36. Klabunde, Anna, and Frans Willekens. 2016. "Decision-making in agent-based models of migration: State of the art and challenges." *European Journal of Population* 32: 73–97. https://doi.org/10.1007/s10680-015-9362-0; Gray, Jonathan, Jason Hilton, and Jakub Bijak. 2017. "Choosing the choice: Reflections on modelling decisions and behaviour in demographic

agent-based models." *Population Studies* 71(1): 85–97. https://doi.org/10.1080/00324728.2017.1350280
37. Koubi, Vally, Gabriele Spilker, Lena Schaffer, and Tobias Böhmelt. 2016. "The role of environmental perceptions in migration decision-making: Evidence from both migrants and non-migrants in five developing countries." *Population and Environment* 38: 134–163. https://doi.org/10.1007/s11111-016-0258-7
38. Khanian, Mojtaba, Behrad Serpoush, and Nima Gheitarani. 2017. "Balance between place attachment and migration based on subjective adaptive capacity in response to climate change: The case of Famenin County in Western Iran." *Climate and Development* 11(1): 69–82. https://doi.org/10.1080/17565529.2017.1374238
39. Partridge, Mark D., Bo Feng, and Mark Rembert. 2017. "Improving climate-change modeling of US migration." *American Economic Review* 107(5): 451–55. https://doi.org/10.1257/aer.p20171054
40. Kahn, Matthew E. 2015. "Climate change adaptation: Lessons from urban economics." *Strategic Behavior and the Environment* 5(1):1–30. http://dx.doi.org/10.1561/102.00000055
41. De Sherbinin, Alex, Kathryn Grace, Sonali McDermid, Kees van der Geest, Michael J. Puma, and Andrew Bell. 2022. "Migration theory in climate mobility research." *Frontiers in Climate* 4: 882343. https://doi.org/10.3389/fclim.2022.882343
42. Hoffmann, Roman, Barbora Šedová, and Kira Vinke. 2021. "Improving the evidence base: A methodological review of the quantitative climate migration literature." *Global Environmental Change* 71: 102367. https://doi.org/10.1016/j.gloenvcha.2021.102367
43. Cattaneo, Cristina, and Giovanni Peri. 2016. "The migration response to increasing temperatures." *Journal of Development Economics* 122: 127–146. https://doi.org/10.1016/j.jdeveco.2016.05.004
44. Cattaneo and Peri 2016.
45. De Haas, Hein. 2021. "A theory of migration: The aspirations-capabilities framework." *Comparative Migration Studies* 9(1): 8. https://doi.org/10.1186/s40878-020-00210-4; Horton, Radley M., Alex de Sherbinin, David Wrathall, and Michael Oppenheimer. 2021. "Assessing human habitability and migration." *Science* 372(6548): 1279–1283. https://doi.org/10.1126/science.abi8603
46. Marchiori, Luca, Jean-François Maystadt, and Ingmar Schumacher. 2017. "Is environmentally induced income variability a driver of human migration?" *Migration and Development* 6(1): 33–59. https://doi.org/10.1080/21632324.2015.1020106
47. Abel, Guy J., Michael Brottrager, Jesus Crespo Cuaresma, and Raya Muttarak. 2019. "Climate, conflict and forced migration." *Global Environmental Change* 54: 239–249. https://doi.org/10.1016/j.gloenvcha.2018.12.003
48. Cottier, Fabien, Marie-Laurence Flahaux, Jesse Ribot, Richard Seager, and Godfreyb Ssekajja. 2022. "Framing the frame: Cause and effect in climate-related migration." *World Development* 158: 106016. https://doi.org/10.1016/j.worlddev.2022.106016
49. Beine, Michel, and Lionel Jeusette. 2021. "A meta-analysis of the literature on climate change and migration." *Journal of Demographic Economics* 87(3): 293–344. https://doi.org/10.1017/dem.2019.22
50. Black, Richard, Nigel W. Arnell, W. Neil Adger, David Thomas, and Andrew Geddes. 2013. "Migration, immobility and displacement outcomes following extreme events." *Environmental Science & Policy* 27: S32–S43. https://doi.org/10.1016/j.envsci.2012.09.001
51. Schutte, S., J. Vestby, J. Carling, and H. Buhaug. 2021. "Climatic conditions are weak predictors of asylum migration." *Nature Communications* 12: 2067. https://doi.org/10.1038/s41467-021-22255-4
52. McLeman, Robert, David Wrathall, Elisabeth Gilmore, Philip Thornton, Helen Adams, and François Gemenne. 2021. "Conceptual framing to link climate risk assessments and climate-migration scholarship." *Climatic Change* 165: 24. https://doi.org/10.1007/s10584-021-03056-6
53. Jones, Bryan, and Alex de Sherbinin. 2022. "Documentation for the groundswell spatial population and migration projections at one-eighth degree according to SSPs and RCPs, 2010-2050." Palisades, NY: NASA Socioeconomic Data and Applications Center (SEDAC). Click here to enter text.
54. Intergovernmental Panel on Climate Change (IPCC). 2022. "Annex II: Glossary," edited by Möller, V., R. van Diemen, J. B. R. Matthews, C. Méndez, S. Semenov, J. S. Fuglestvedt, and

A. Reisinger. In *Climate Change 2022: Impacts, Adaptation and Vulnerability. Contribution of Working Group II to the Sixth Assessment Report of the Intergovernmental Panel on Climate Change*, edited by Pörtner, H.-O., D. C. Roberts, M. Tignor, E. S. Poloczanska, K. Mintenbeck, A. Alegría, M. Craig, S. Langsdorf, S. Löschke, V. Möller, A. Okem, and B. Rama, 2897–2930. Cambridge: Cambridge University Press. https://doi.org/10.1017/9781009325844.029

55. IPCC 2022. Note: SSPs are also integrated with RCPs to the extent that SSP1 through SSP5 may be denoted as SSP1-1.9 to SSP5-8.5.
56. Best, Kelsea, Jonathan Gilligan, Hiba Baroud, Amanda Carrico, Katharine Donato, and Bishawjit Mallick. 2022. "Applying machine learning to social datasets: A study of migration in southwestern Bangladesh using random forests." *Regional Environmental Change* 22(2): 52. https://doi.org/10.1007/s10113-022-01915-1
57. Preston, Benjamin L., Johanna Mustelin, and Megan C. Maloney. 2015. "Climate adaptation heuristics and the science/policy divide." *Mitigation and Adaptation Strategies for Global Change* 20: 467–497. https://doi.org/10.1007/s11027-013-9503-x
58. Beyer, Robert, and Milan, Andrea. 2023. *Climate Change and Human Mobility: Quantitative Evidence on Global Historical Trends and Future Projections*. Berlin: Global Data Institute/International Organization for Migration. https://gmdac.iom.int/sites/g/files/tmzbdl1416/files/documents/2023-06/final_2023-climate-change-and-human-mobility.pdf
59. Schewel, Kerilyn, Sarah Dickerson, B. Madson, and Gabriela Nagle Alverio. 2024. "How well can we predict climate migration? A review of forecasting models." *Frontiers in Climate* 5: 1189125. https://doi.org/10.3389/fclim.2023.1189125
60. Poston, Dudley L., Jr., Li Zhang, David J. Gotcher, and Yuan Gu. 2009. "The effect of climate on migration: United States, 1995–2000." *Social Science Research* 38(3): 743–753. https://doi.org/10.1016/j.ssresearch.2008.10.003
61. Fussell, Elizabeth, and Brianna Castro. 2022. "Environmentally informed migration in North America." In *International Handbook of Population and Environment*, edited by Lori M. Hunter, Clark Gray, and Jacques Veron, 205–223. Cham: Springer. https://doi.org/10.1007/978-3-030-76433-3_10
62. Poston et al. 2009.
63. Glaeser, Edward L., and Kristina Tobio. 2008. "The rise of the Sunbelt." *Southern Economic Journal* 74(3): 609–643. https://doi.org/10.1002/j.2325-8012.2008.tb00856.x
64. Glaeser and Tobio 2008.
65. Jewell, Katherine R. 2020. "The rise of the Sunbelt South." In *Oxford Research Encyclopedia of American History*. Oxford: Oxford University Press. https://doi.org/10.1093/acrefore/9780199329175.013.833
66. Rappaport, Jordan. 2007. "Moving to nice weather." *Regional Science and Urban Economics* 37(3): 375–398. https://doi.org/10.1016/j.regsciurbeco.2006.11.004
67. Barreca, Alan, Karen Clay, Olivier Deschenes, Michael Greenstone, and Joseph S. Shapiro. 2016. "Adapting to climate change: The remarkable decline in the US temperature-mortality relationship over the twentieth century." *Journal of Political Economy* 124(1): 105–159. https://doi.org/10.1086/684582
68. Horton, Radley M., Alex de Sherbinin, David Wrathall, and Michael Oppenheimer. 2021. "Assessing human habitability and migration." *Science* 372(6548): 1279–1283. https://doi.org/10.1126/science.abi8603
69. Rappaport 2007.
70. Sinha, Paramita, Martha L. Caulkins, and Maureen L. Cropper. 2018. "Household location decisions and the value of climate amenities." *Journal of Environmental Economics and Management* 92: 608–637. https://doi.org/10.1016/j.jeem.2017.08.005
71. Sinha et al. 2018.
72. Leduc, Sylvain, and Daniel J. Wilson. 2024. "Snow Belt to Sun Belt migration: End of an era?" Working Paper 2024-21. San Francisco: Federal Reserve Bank of San Francisco. https://doi.org/10.24148/wp2024-21
73. Leduc and Wilson 2024.
74. Leduc and Wilson 2024.
75. Samarin, Mikhail, and Madhuri Sharma. 2023. "A typology of US metropolises by rent burden and its major drivers." *GeoJournal* 88(5): 4887–4906. https://doi.org/10.1007/s10708-023-10898-3

76. Scheier, Eric, and Noah Kittner. 2022. "A measurement strategy to address disparities across household energy burdens." *Nature Communications* 13(1): 288. https://doi.org/10.1038/s41467-021-27673-y
77. Ross, Andrew. 2021. *Sunbelt Blues: The Failure of American Housing*. New York: Metropolitan Books.
78. Wrenn, Douglas H. 2024. "The effect of natural disasters and extreme weather on household location choice and economic welfare." *Journal of the Association of Environmental and Resource Economists* 11(5): 1101–1134. https://doi.org/10.1086/728887
79. Chen, I-Ching, Jane K. Hill, Ralf Ohlemüller, David B. Roy, and Chris D. Thomas. 2011. "Rapid range shifts of species associated with high levels of climate warming." *Science* 333 (6045): 1024–1026. https://doi.org/10.1126/science.1206432
80. Beaury, Evelyn M., Emily J. Fusco, Michelle R. Jackson, Brittany B. Laginhas, Toni Lyn Morelli, Jenica M. Allen, Valerie J. Pasquarella, and Bethany A. Bradley. 2020. "Incorporating climate change into invasive species management: Insights from managers." *Biological Invasions* 22: 233–252. https://doi.org/10.1007/s10530-019-02087-6
81. Dale, Virginia H., Linda A. Joyce, Steve McNulty, Ronald P. Neilson, Matthew P. Ayres, Michael D. Flannigan, Paul J. Hanson, Llyod C. Irland, Ariel E. Lugo, and Chris J. Peterson. 2001. "Climate change and forest disturbances: Climate change can affect forests by altering the frequency, intensity, duration, and timing of fire, drought, introduced species, insect and pathogen outbreaks, hurricanes, windstorms, ice storms, or landslides." *BioScience* 51(9): 723–734. https://doi.org/10.1641/0006-3568(2001)051[0723:CCAFD]2.0.CO;2
82. Ward, Philip J., Marleen C. de Ruiter, Johanna Mård, Kai Schröter, Anne Van Loon, Ted Veldkamp, Nina von Uexkull, and Niko Wanders. 2020. "The need to integrate flood and drought disaster risk reduction strategies." *Water Security* 11: 100070. https://doi.org/10.1016/j.wasec.2020.100070
83. Carlson, Colin J., Gregory F. Albery, Cory Merow, Christopher H. Trisos, Casey M. Zipfel, Evan A. Eskew, Kevin J. Olival, Noam Ross, and Shweta Bansal. 2022. "Climate change increases cross-species viral transmission risk." *Nature* 607 (7919): 555–562. https://doi.org/10.1038/s41586-022-04788-w
84. Semenza, Jan C., and Kristie L. Ebi. 2019. "Climate change impact on migration, travel, travel destinations and the tourism industry." *Journal of Travel Medicine* 26(5): taz026. https://doi.org/10.1093/jtm/taz026
85. Herrera, Clary, Alexis U. Nkusi, Emaline Laney, Morgan A. Lane, Amitha Sampath, Divya R. Bhamidipati, Uriel Kitron, Rebecca Philipsborn, Cassandra White, and Jessica K. Fairley. 2024. "Climate drivers of migration and neglected tropical disease burden in Latin American and Caribbean immigrants: A pilot study in Atlanta, Georgia." *The Journal of Climate Change and Health* 17: 100308. https://doi.org/10.1016/j.joclim.2024.100308
86. Sikk, Kaarel, and Geoffrey Caruso. 2024. "Framing settlement systems as spatial adaptive systems." *Ecological Modelling* 490: 110652. https://doi.org/10.1016/j.ecolmodel.2024.110652
87. Ran, Min, and Liang Chen. 2019. "The 4.2 ka BP climatic event and its cultural responses." *Quaternary International* 521: 158–167. https://doi.org/10.1016/j.quaint.2019.05.030
88. Brook, Timothy. 2023. *The Price of Collapse: The Little Ice Age and the Fall of Ming China*. Princeton, NJ: Princeton University Press.
89. Middleton, Guy D. 2025. "Collapse studies in archaeology from 2012 to 2023." *Journal of Archaeological Research* 33: 57–115. https://doi.org/10.1007/s10814-024-09196-4
90. Van de Noort, Robert. 2013. *Climate Change Archaeology: Building Resilience from Research in the World's Coastal Wetlands*. New York: Oxford University Press.
91. Middleton 2025.
92. Johnson, Scott A. J. 2016. *Why Did Ancient Civilizations Fail?* New York: Routledge.
93. Orr, Scott Allan, Jenny Richards, and Sandra Fatorić. 2022. "Climate change and cultural heritage: A systematic literature review (2016–2020)." *The Historic Environment: Policy & Practice* 12(3–4): 434–477. https://doi.org/10.1080/17567505.2021.1957264
94. Dembedza, Vimbainashe Prisca, Prosper Chopera, Jacob Mapara, and Lesley Macheka. 2022. "Impact of climate change-induced natural disasters on intangible cultural heritage related to food: A review." *Journal of Ethnic Foods* 9(1): 32. https://doi.org/10.1186/s42779-022-00147-2

95. Heyd, Thomas. 2013. "Landmarks of the sacred in times of climate change." In *Aesth/ethics in Environmental Change: Hiking Through the Arts, Ecology, Religion and Ethics of the Environment*, edited by Sigurd Bergmann, Irmgard Blindow, and Konrad Ott, 143–158. Zurich: Verlag Lit.
96. Butzer, Karl W. 2012. "Collapse, environment, and society." *Proceedings of the National Academy of Sciences* 109(10): 3632–3639. https://doi.org/10.1073/pnas.1114845109
97. Butzer 2012.

Chapter 2

1. Wheeler, Stephen M. 2015. "Built landscapes of metropolitan regions: An international typology." *Journal of the American Planning Association* 81(3): 167–190. https://doi.org/10.1080/01944363.2015.1081567
2. Moffatt, Sebastian, and Niklaus Kohler. 2008. "Conceptualizing the built environment as a social-ecological system." *Building Research & Information* 36(3): 248–268. https://doi.org/10.1080/09613210801928131
3. Forman, Richard T. T. 2016. "Urban ecology principles: Are urban ecology and natural area ecology really different?" *Landscape Ecology* 31: 1653–1662. https://doi.org/10.1007/s10980-016-0424-4
4. Opoku, Alex. 2019. "Biodiversity and the built environment: Implications for the Sustainable Development Goals (SDGs)." *Resources, Conservation and Recycling* 141: 1–7. https://doi.org/10.1016/j.resconrec.2018.10.011
5. P. R. Shukla, J. Skea, R. Slade, R. van Diemen, E. Haughey, J. Malley, M. Pathak, J. Portugal Pereira, eds. 2019. "Technical summary." In *Climate Change and Land: An IPCC special report on climate change, desertification, land degradation, sustainable land management, food security, and greenhouse gas fluxes in terrestrial ecosystems*, edited by Shukla, P. R., J. Skea, E. Calvo Buendia, V. Masson-Delmotte, H.-O. Pörtner, D. C. Roberts, P. Zhai, R. Slade, S. Connors, R. van Diemen, M. Ferrat, E. Haughey, S. Luz, S. Neogi, M. Pathak, J. Petzold, J. Portugal Pereira, P. Vyas, E. Huntley, K. Kissick, M, Belkacemi, and J. Malley. https://doi.org/10.1017/9781009157988.002
6. Shukla et al. 2019.
7. Kress, W. John, and Jeffrey K. Stine, eds. 2017. *Living in the Anthropocene: Earth in the Age of Humans*. Washington, D.C.: Smithsonian Institution.
8. Devlin, Ann Sloan. 2018. *Environmental Psychology and Human Well-being: Effects of Built and Natural Settings*. London: Academic Press.
9. Brady, Emily. 2019. *Aesthetics of the Natural Environment*. Edinburgh: Edinburgh University Press.
10. McClelland, Linda Flint. 1998. *Building the National Parks: Historic Landscape Design and Construction*. Baltimore: John Hopkins University Press.
11. Ness, David A., and Ke Xing. 2017. "Toward a resource-efficient built environment: A literature review and conceptual model." *Journal of Industrial Ecology* 21(3): 572–592. https://doi.org/10.1111/jiec.12586
12. Federal Reserve Bank of St. Louis. 2024. "FRED economic data: Housing inventory estimate; Total housing units in the United States." St. Louis, MO: Federal Reserve Bank of St. Louis. Retrieved from https://fred.stlouisfed.org/series/ETOTALUSQ176N
13. U.S. Energy Information Administration. 2024. "How much energy is consumed in U.S. buildings?" Washington, D.C.: U.S. Department of Energy. https://www.eia.gov/tools/faqs/faq.php?id=86&t=1
14. U.S. Energy Information Administration. 2018. "2018 commercial building energy consumption survey: Building characteristics highlights." Washington, D.C.: U.S. Department of Energy. https://www.eia.gov/consumption/commercial/data/2018/pdf/CBECS_2018_Building_Characteristics_Flipbook.pdf
15. National Institute of Standards and Technology (NIST). 2023. "The manufacturing economy." Gaithersburg, MD: U.S. Department of Commerce. https://www.nist.gov/el/applied-economics-office/manufacturing/manufacturing-economy
16. U.S. Department of Energy. 2022. *Industrial Decarbonization Roadmap*. Washington, D.C.: U.S. Department of Energy. https://www.energy.gov/sites/default/files/2022-09/Industrial%20Decarbonization%20Roadmap.pdf

17. Author's calculations based on Food and Agriculture Organization of the United Nations. 2023. *World Food and Agriculture—Statistical Yearbook 2022.* Rome: United Nations. Retrieved from https://openknowledge.fao.org/handle/20.500.14283/cc2211en
18. U.S. Energy Information Administration. 2024. "How much energy does a person use in a year?" Washington, D.C.: U.S. Department of Energy. https://www.eia.gov/tools/faqs/faq.php?id=85&t=1
19. World Population Review. 2024. "Energy consumption by country 2024." Retrieved from https://worldpopulationreview.com/country-rankings/energy-consumption-by-country
20. World Bank. 2022. "CO_2 emissions (metric tons per capita)" Washington, D.C.: World Bank Group. https://databank.worldbank.org/metadataglossary/world-development-indicators/series/EN.ATM.CO2E.PC
21. International Energy Agency. 2023. *Breakthrough Agenda Report 2023: Buildings.* Paris: International Energy Agency. https://www.iea.org/reports/breakthrough-agenda-report-2023/buildings
22. Cabeza, L. F., Q. Bai, P. Bertoldi, J. M. Kihila, A. F. P. Lucena, É. Mata, S. Mirasgedis, A. Novikova, and Y. Saheb. 2022. "Buildings." In IPCC, *Climate Change 2022.*
23. Chu, E. K., M. M. Fry, J. Chakraborty, S.-M. Cheong, C. Clavin, M. Coffman, D. M. Hondula, D. Hsu, V. L. Jennings, J. M. Keenan, A. Kosmal, T. A. Muñoz-Erickson, and N. T. O. Jelks. 2023. "Ch. 12: Built environment, urban systems, and cities." In *Fifth National Climate Assessment,* edited by Crimmins, A. R., C. W. Avery, D. R. Easterling, K. E. Kunkel, B. C. Stewart, and T. K. Maycock. Washington, D.C.: U.S. Global Change Research Program. https://doi.org/10.7930/NCA5.2023.CH12
24. Chu et al. 2023.
25. U.S. Census Bureau. 2023. "2020 census urban area facts." Washington, D.C.: U.S. Census Bureau. https://www.census.gov/programs-surveys/geography/guidance/geo-areas/urban-rural/2020-ua-facts.html
26. U.S. Census Bureau. 2012. "Increasing urbanization: Population distribution by city size, 1790 to 1890." Washington, D.C.: U.S. Census Bureau. https://www.census.gov/dataviz/visualizations/005/
27. Krieger, Alex. 2019. *City on a Hill: Urban Idealism in America from the Puritans to the Present.* Cambridge, MA: Harvard University Press.
28. World Bank. 2024. "Understanding poverty: Urban development." Washington, D.C.: World Bank Group. https://www.worldbank.org/en/topic/urbandevelopment/overview
29. Blaine, James G., Bernard W. Sweeney, and David B. Arscott. 2006. "Enhanced source-water monitoring for New York City: Historical framework, political context, and project design." *Journal of the North American Benthological Society* 25(4): 851–866.
30. Solecki, William D., and Cynthia Rosenzweig. 2006. "Biodiversity, biosphere reserves, and the Big Apple: a study of the New York Metropolitan Region." *Annals of the New York Academy of Sciences* 1023(1): 105–124. https://doi.org/10.1196/annals.1319.004
31. Rees, William E. and Wackernagel, Mathis, 1996. "Urban ecological footprints: Why cities cannot be sustainable and why they are a key to sustainability." *Environmental Impact Assessment Review* 16(4–6), 223–248.
32. Elhacham, Emily, Liad Ben-Uri, Jonathan Grozovski, Yinon M. Bar-On, and Ron Milo. 2020. "Global human-made mass exceeds all living biomass." *Nature* 588(7838): 442–444. https://doi.org/10.1038/s41586-020-3010-5
33. Busch, Pablo, Alissa Kendall, Colin W. Murphy, and Sabbie A. Miller. 2022. "Literature review on policies to mitigate GHG emissions for cement and concrete." *Resources, Conservation and Recycling* 182: 106278. https://doi.org/10.1016/j.resconrec.2022.106278
34. Center for Sustainable Systems. 2024. "U.S. environmental footprint factsheet." Ann Arbor: University of Michigan. https://css.umich.edu/publications/factsheets/sustainability-indicators/us-environmental-footprint-factsheet
35. Environmental Protection Agency (EPA). 2023. "Construction and demolition debris: Material-specific data." Washington, D.C.: Environmental Protection Agency. https://www.epa.gov/facts-and-figures-about-materials-waste-and-recycling/construction-and-demolition-debris-material
36. Author's calculations based on Yuan, Liang, Bing Yang, Weisheng Lu, and Ziyu Peng. 2024. "Carbon footprint accounting across the construction waste lifecycle: A critical review of

research." *Environmental Impact Assessment Review* 107: 107551. https://doi.org/10.1016/j.eiar.2024.107551

37. Environmental Protection Agency (EPA). 2024. "Greenhouse gas equivalencies calculator." Washington, D.C.: Environmental Protection Agency.
38. Purchase, Callun Keith, Dhafer Manna Al Zulayq, Bio Talakatoa O'Brien, Matthew Joseph Kowalewski, Aydin Berenjian, Amir Hossein Tarighaleslami, and Mostafa Seifan. 2022. "Circular economy of construction and demolition waste: A literature review on lessons, challenges, and benefits." *Materials* 15(1): 76. https://doi.org/10.3390/ma15010076
39. Sormunen, Petri, and Timo Kärki. 2019. "Recycled construction and demolition waste as a possible source of materials for composite manufacturing." *Journal of Building Engineering* 24: 100742.
40. Blair, John, and Sarath Mataraarachchi. 2021. "A review of landfills, waste and the nearly forgotten nexus with climate change." *Environments* 8(8): 73. https://doi.org/10.3390/environments8080073
41. Hardy, Steve. 2017, July 5th. "Rising sea levels threaten Louisiana homeowners with billions in losses, real estate economists say." *The Times-Picayune / The New Orleans Advocate.* https://www.nola.com/rising-sea-levels-threaten-louisiana-homeowners-with-billions-in-losses-real-estate-economists-say/article_f7ed3deb-6c53-5a89-a779-cf0cc5dd108b.html
42. Author's calculations based on Municipal and Industrial Solid Waste Division, Office of Solid Waste, Environmental Protection Agency. 1998. *Characterization of Building-Related Construction and Demolition Debris in the United States.* Report No. EPA530-R-98-010. Washington, D.C.: Environmental Protection Agency; Federal Reserve Bank of St. Louis. 2024. "FRED economic data: Housing inventory; Median home size in square feet in Louisiana." St. Louis, MO: Federal Reserve Bank of St. Louis. https://fred.stlouisfed.org/series/MEDSQUFEELA. Average existing housing inventory size is 1,800 ft2.
43. Merschroth, Simon, Alessio Miatto, Steffi Weyand, Hiroki Tanikawa, and Liselotte Schebek. 2020. "Lost material stock in buildings due to sea level rise from global warming: The case of Fiji Islands." *Sustainability* 12(3): 834. https://doi.org/10.3390/su12030834
44. Islam, Rashidul, Tasnia Hassan Nazifa, Adhi Yuniarto, A. S. M. Shanawaz Uddin, Salmiati Salmiati, and Shamsuddin Shahid. 2019. "An empirical study of construction and demolition waste generation and implication of recycling." *Waste Management* 95: 10–21. https://doi.org/10.1016/j.wasman.2019.05.049
45. Beaven, R. P., A. M. Stringfellow, R. J. Nicholls, I. D. Haigh, Abiy S. Kebede, and Jenny Watts. 2020. "Future challenges of coastal landfills exacerbated by sea level rise." *Waste Management* 105: 92–101. https://doi.org/10.1016/j.wasman.2020.01.027
46. Stafford, Sarah L., and Alexander D. Renaud. 2019. "Measuring the potential for toxic exposure from storm surge and sea-level rise: Analysis of coastal Virginia." *Natural Hazards Review* 20(1): 04018024. https://doi.org/10.1061/(ASCE)NH.1527-6996.0000315
47. Cushing, Lara J., Yang Ju, Scott Kulp, Nicholas Depsky, Seigi Karasaki, Jessie Jaeger, Amee Raval, Benjamin Strauss, and Rachel Morello-Frosch. 2023. "Toxic tides and environmental injustice: Social vulnerability to sea level rise and flooding of hazardous sites in coastal California." *Environmental Science & Technology* 57(19): 7370–7381. https://doi.org/10.1021/acs.est.2c07481
48. Adam, Issahaku. 2021. "Tourists' perception of beach litter and willingness to participate in beach clean-up." *Marine Pollution Bulletin* 170: 112591. https://doi.org/10.1016/j.marpolbul.2021.112591
49. Pendleton, Linwood, Daniel C. Donato, Brian C. Murray, Stephen Crooks, W. Aaron Jenkins, Samantha Sifleet, and Christopher Craft. 2012. "Estimating global 'blue carbon' emissions from conversion and degradation of vegetated coastal ecosystems." *PLOS One*: e43542. https://doi.org/10.1371/journal.pone.0043542
50. Berrill, Peter, Kenneth T. Gillingham, and Edgar G. Hertwich. 2021. "Linking housing policy, housing typology, and residential energy demand in the United States." *Environmental Science & Technology* 55(4): 2224–2233. https://doi.org/10.1021/acs.est.0c05696
51. Burd, Charlynn, Michael Burrows, and Brian McKenzie. 2021. *Travel Time to Work in the United States: 2019.* American Community Survey Reports. Washington, D.C.: United States Census Bureau.
52. Bereitschaft, Bradley. 2023. "Do socially vulnerable urban populations have access to walkable, transit-accessible neighborhoods? A nationwide analysis of large US metropolitan areas." *Urban Science* 7(1): 6. https://doi.org/10.3390/urbansci7010006

53. St. John, Jeff. 2024. "More demand, more gas: Inside the Southeast's dirty power push." *Canary Media*. https://www.canarymedia.com/articles/utilities/more-demand-more-gas-inside-the-southeasts-dirty-power-push
54. Goldstein, Benjamin, Dimitrios Gounaridis, and Joshua P. Newell. 2020. "The carbon footprint of household energy use in the United States." *Proceedings of the National Academy of Sciences* 117(32): 19122–19130. https://doi.org/10.1073/pnas.1922205117
55. Missimer, Thomas M., Philip A. Danser, Gary Amy, and Thomas Pankratz. 2014. "Water crisis: The metropolitan Atlanta, Georgia, regional water supply conflict." *Water Policy* 16(4): 669–689. https://doi.org/10.2166/wp.2014.131
56. Jones, Christopher, and Daniel M. Kammen. 2014. "Spatial distribution of US household carbon footprints reveals suburbanization undermines greenhouse gas benefits of urban population density." *Environmental Science & Technology* 48(2): 895–902. https://doi.org/10.1021/es4034364
57. Muñiz, Ivan, and Andrés Dominguez. 2020. "The impact of urban form and spatial structure on per capita carbon footprint in US larger metropolitan areas." *Sustainability* 12(1): 389. https://doi.org/10.3390/su12010389
58. Interview with South Florida Mayor, conducted in-person by the author, January 27, 2017.
59. Pinto, F. Silva, B. de Carvalho, and R. Cunha Marques. 2021. "Adapting water tariffs to climate change: Linking resource availability, costs, demand, and tariff design flexibility." *Journal of Cleaner Production* 290: 125803. https://doi.org/10.1016/j.jclepro.2021.125803; Barreca, Alan, R. Jisung Park, and Paul Stainier. 2022. "High temperatures and electricity disconnections for low-income homes in California." *Nature Energy* 7 (11): 1052–1064. https://doi.org/10.1038/s41560-022-01134-2
60. Keenan, Jesse M. 2018. "Types and forms of resilience in local planning in the US: Who does what?" *Environmental Science & Policy* 88: 116–123. https://doi.org/10.1016/j.envsci.2018.06.015
61. Shi, Linda, and Susanne Moser. "Transformative climate adaptation in the United States: Trends and prospects." *Science* 372 (6549): eabc8054. https://doi.org/10.1126/science.abc8054
62. Kunkel, Kenneth E., Thomas R. Karl, Michael F. Squires, Xungang Yin, Steve T. Stegall, and David R. Easterling. 2020. "Precipitation extremes: Trends and relationships with average precipitation and precipitable water in the contiguous United States." *Journal of Applied Meteorology and Climatology* 59(1): 125–142. https://doi.org/10.1175/JAMC-D-19-0185.1; Kirchmeier-Young, Megan C., and Xuebin Zhang. 2020. "Human influence has intensified extreme precipitation in North America." *Proceedings of the National Academy of Sciences* 117(24): 13308–13313. https://doi.org/10.1073/pnas.1921628117
63. Akinsanola, A. Akintomide, Gabriel J. Kooperman, Kevin A. Reed, Angeline G. Pendergrass, and Walter M. Hannah. 2020. "Projected changes in seasonal precipitation extremes over the United States in CMIP6 simulations." *Environmental Research Letters* 15(10): 104078. https://doi.org/10.1088/1748-9326/abb397
64. National Aeronautics and Space Administration (NASA). 2022. *Steamy Relationships: How Atmospheric Water Vapor Amplifies Earth's Greenhouse Effect*. Washington, D.C.: National Aeronautics and Space Administration.
65. National Weather Services. 2022. "Flood related hazards." Silver Spring, MD: National Oceanographic and Atmospheric Administration. https://www.weather.gov/safety/flood-hazards
66. Centers for Disease Control and Prevention (CDC). 2024. "Precipitation extremes." Atlanta: Centers for Disease Control and Prevention. https://www.cdc.gov/climate-health/php/effects/precipitation-extremes.html
67. Tellman, Beth, Cody Schank, Bessie Schwarz, Peter D. Howe, and Alex de Sherbinin. 2020. "Using disaster outcomes to validate components of social vulnerability to floods: Flood deaths and property damage across the USA." *Sustainability* 12(15): 6006. https://doi.org/10.3390/su12156006
68. Anderson, Curt. 2024, June 13th. "What to know about a series of storms that has swamped South Florida with flash floods." Associated Press. https://apnews.com/article/rain-florida-flooding-what-to-know-6f0a512f1406cee0e37a31df1c75d3cf
69. National Weather Services. 2018. "Hurricane Harvey." Silver Spring, MD: National Oceanographic and Atmospheric Administration. https://www.weather.gov/hgx/hurricaneharvey
70. Risser, Mark D., and Michael F. Wehner. 2017. "Attributable human-induced changes in the likelihood and magnitude of the observed extreme precipitation during Hurricane Harvey." *Geophysical Research Letters* 44(24): 12–457. https://doi.org/10.1002/2017GL075888

71. National Centers for Environmental Information. 2024. "Climate monitoring: National trends; Average annual precipitation trends." Asheville, NC: National Oceanographic and Atmospheric Administration. https://www.ncei.noaa.gov/access/monitoring/us-trends/prcp/ann
72. Marvel, K., W. Su, R. Delgado, S. Aarons, A. Chatterjee, M. E. Garcia, Z. Hausfather, K. Hayhoe, D. A. Hence, E. B. Jewett, A. Robel, D. Singh, A. Tripati, and R. S. Vose. 2023. "Climate trends." In Crimmins et al., *Fifth National Climate Assessment.*
73. Marvel et al. 2023.
74. National Centers for Environmental Information. 2024. State Climate Extremes Committee: Records. Asheville, NC: National Oceanographic and Atmospheric Administration. https://www.ncei.noaa.gov/access/monitoring/scec/records/all/maxp
75. Raupach, Timothy H., Olivia Martius, John T. Allen, Michael Kunz, Sonia Lasher-Trapp, Susanna Mohr, Kristen L. Rasmussen, Robert J. Trapp, and Qinghong Zhang. 2021. "The effects of climate change on hailstorms." *Nature Reviews Earth & Environment* 2(3): 213–226. https://doi.org/10.1038/s43017-020-00133-9
76. Carrns, Ann. 2024, March 22nd. "Insurers report rising hail damage claims." *The New York Times.* https://www.nytimes.com/2024/03/22/your-money/insurers-report-rising-hail-damage-claims.html
77. CAPE Analytics. 2024. "Hail risk: The growing threat for property insurers." https://capeanalytics.com/blog/hail-risk-for-property-insurers/
78. National Centers for Environmental Information. 2024. "Billion-dollar weather and climate disasters: 2021." Asheville, NC: National Oceanographic and Atmospheric Administration. https://www.ncei.noaa.gov/access/billions/events/US/2021
79. Hohl, Roman, Hans-Heinrich Schiesser, and Dörte Aller. 2002. "Hailfall: The relationship between radar-derived hail kinetic energy and hail damage to buildings." *Atmospheric Research* 63 (3–4): 177–207. https://doi.org/10.1016/S0169-8095(02)00059-5
80. Klima, Kelly, and M. Granger Morgan. 2015. "Ice storm frequencies in a warmer climate." *Climatic Change* 133: 209–222. https://doi.org/10.1007/s10584-015-1460-9
81. Houston, Tamara G., and Stanley A. Changnon. 2007. "Freezing rain events: A major weather hazard in the conterminous US." *Natural Hazards* 40(2): 485–494. https://doi.org/10.1007/s11069-006-9006-0
82. Gu, Yaofeng, and Dan Li. 2018. "A modeling study of the sensitivity of urban heat islands to precipitation at climate scales." *Urban Climate* 24: 982–993. https://doi.org/10.1016/j.uclim.2017.12.001
83. Hintz, William D., Laura Fay, and Rick A. Relyea. 2022. "Road salts, human safety, and the rising salinity of our fresh waters." *Frontiers in Ecology and the Environment* 20(1): 22–30. https://doi.org/10.1002/fee.2433
84. Kaushal, Sujay S. 2016. "Increased salinization decreases safe drinking water." *Environmental Science and Technology* 50(6): 2765–2766. https://doi.org/10.1021/acs.est.6b00679
85. Marshall, Adrienne M., John T. Abatzoglou, Stefan Rahimi, Dennis P. Lettenmaier, and Alex Hall. 2024. "California's 2023 snow deluge: Contextualizing an extreme snow year against future climate change." *Proceedings of the National Academy of Sciences* 121(20): e2320600121. https://doi.org/10.1073/pnas.2320600121
86. O'Gorman, Paul A. 2014. "Contrasting responses of mean and extreme snowfall to climate change." *Nature* 512(7515): 416–418. https://doi.org/10.1038/nature13625; Quante, Lennart, Sven N. Willner, Robin Middelanis, and Anders Levermann. 2021. "Regions of intensification of extreme snowfall under future warming." *Scientific Reports* 11(1): 16621. https://doi.org/10.1038/s41598-021-95979-4
87. Croce, Pietro, Paolo Formichi, Filippo Landi, and Francesca Marsili. 2018. "Climate change: Impact on snow loads on structures." *Cold Regions Science and Technology* 150: 35–50. https://doi.org/10.1016/j.coldregions.2017.10.009
88. Arctic Design Group. 2024. "About the Arctic Design Group." Charlottesville: University of Virginia, School of Architecture. https://arch.virginia.edu/research/arctic-design-group
89. Lamichhane, Jay Ram. 2021. "Rising risks of late-spring frosts in a changing climate." *Nature Climate Change* 11(7): 554–555. https://doi.org/10.1038/s41558-021-01090-x
90. Van Aarle, Marcel, Henk Schellen, and Jos van Schijndel. 2015. "Hygro thermal simulation to predict the risk of frost damage in masonry; effects of climate change." *Energy Procedia* 78: 2536–2541. https://doi.org/10.1016/j.egypro.2015.11.268

91. New York City Mayor's Office of Climate & Environmental Justice. 2022. *Climate Resiliency Design Guidelines*. New York: City of New York. https://www.nyc.gov/assets/sustainability/downloads/pdf/publications/CRDG-4-1-May-2022.pdf
92. Cook, Lauren M., Christopher J. Anderson, and Constantine Samaras. 2017. "Framework for incorporating downscaled climate output into existing engineering methods: Application to precipitation frequency curves." *Journal of Infrastructure Systems* 23(4): 04017027. https://doi.org/10.1061/(ASCE)IS.1943-555X.0000382
93. Olsen, J. Rolf, ed. 2015. *Adapting Infrastructure and Civil Engineering Practice to a Changing Climate*. Reston, VA: American Society of Civil Engineers, Committee on Adaptation to a Changing Climate. http://dx.doi.org/10.1061/9780784479193
94. Pralle, Sarah. 2019. "Drawing lines: FEMA and the politics of mapping flood zones." *Climatic Change* 152(2): 227–237. https://doi.org/10.1007/s10584-018-2287-y
95. Herreros-Cantis, Pablo, Veronica Olivotto, Zbigniew J. Grabowski, and Timon McPhearson. 2020. "Shifting landscapes of coastal flood risk: Environmental (in) justice of urban change, sea level rise, and differential vulnerability in New York City." *Urban Transformations* 2(1): 9. https://doi.org/10.1186/s42854-020-00014-w
96. Nadal, Norberto C., Raúl E. Zapata, Ismael Pagán, Ricardo López, and Jairo Agudelo. 2010. "Building damage due to riverine and coastal floods." *Journal of Water Resources Planning and Management* 136(3): 327–336. https://doi.org/10.1061/(ASCE)WR.1943-5452.0000036
97. Bikçe, Murat, Murat Örnek, and Ömer Faruk Cansız. 2018. "The effect of buoyancy force on structural damage: A case study." *Engineering Failure Analysis* 92: 553–565. https://doi.org/10.1016/j.engfailanal.2018.06.014
98. Xiao, Jianzhuang, Chengbing Qiang, Antonio Nanni, and Kaijian Zhang. 2017. "Use of sea-sand and seawater in concrete construction: Current status and future opportunities." *Construction and Building Materials* 155: 1101–1111. https://doi.org/10.1016/j.conbuildmat.2017.08.130
99. Ho, Michelle, Upmanu Lall, Maura Allaire, Naresh Devineni, Hyun Han Kwon, Indrani Pal, David Raff, and David Wegner. 2017. "The future role of dams in the United States of America." *Water Resources Research* 53(2): 982–998. https://doi.org/10.1002/2016WR019905
100. Gonzales, Vincent, and Margaret Walls. 2020. *Dams and Dam Removals in the United States*. Washington, D.C.: Resources for the Future. https://media.rff.org/documents/Dams_and_Dam_Removals_in_the_US_Js8EO7S.pdf
101. Smith, Mitch. 20204, June 25th. "How the Midwest floods nearly took out a century-old dam." *The New York Times*. https://www.nytimes.com/2024/06/25/us/minnesota-dam-flooding.html?
102. Rick, B., D. McGrath, S. W. McCoy, and W. H. Armstrong. 2023. "Unchanged frequency and decreasing magnitude of outbursts from ice-dammed lakes in Alaska." *Nature Communications* 14(1): 6138. https://doi.org/10.1038/s41467-023-41794-6
103. Florida Department of Environmental Protection. 2023. Geospatial Open Data: Florida Dams. Tallahassee: State of Florida. https://geodata.dep.state.fl.us/datasets/FDEP::florida-dams/about
104. Brody, Samuel D., Sammy Zahran, Praveen Maghelal, Himanshu Grover, and Wesley E. Highfield. 2007. "The rising costs of floods: Examining the impact of planning and development decisions on property damage in Florida." *Journal of the American Planning Association* 73(3): 330–345. https://doi.org/10.1080/01944360708977981
105. Seltenrich, Nate. 2018. "Safe from the storm: Creating climate-resilient health care facilities." *Environmental Health Perspectives* 126(10): 102001. https://doi.org/10.1289/EHP3810
106. Salas, Renee N., Laura G. Burke, Jessica Phelan, Gregory A. Wellenius, E. John Orav, and Ashish K. Jha. 2024. "Impact of extreme weather events on healthcare utilization and mortality in the United States." *Nature Medicine* 30: 1118–1126. https://doi.org/10.1038/s41591-024-02833-x
107. Sun, Pin, Rebecca Entress, Jenna Tyler, Abdul-Akeem Sadiq, and Douglas Noonan. 2023. "Critical public infrastructure underwater: The flood hazard profile of Florida hospitals." *Natural Hazards* 117(1): 473–489. https://doi.org/10.1007/s11069-023-05869-3
108. Rhoades, Jason L., James S. Gruber, and Bill Horton. 2018. "Developing an in-depth understanding of elderly adult's vulnerability to climate change." *The Gerontologist* 58(3): 567–577. https://doi.org/10.1093/geront/gnw167

109. Qiang, Yi. "Disparities of population exposed to flood hazards in the United States." *Journal of Environmental Management* 232: 295–304. https://doi.org/10.1016/j.jenvman.2018.11.039
110. Braveman, Paula A., Catherine Cubbin, Susan Egerter, David R. Williams, and Elsie Pamuk. 2011. "Socioeconomic disparities in health in the United States: What the patterns tell us." *American Journal of Public Health* 100 (S1): S186–S196. https://doi.org/10.2105/AJPH.2009.166082; Smith, Genee S., E. Anjum, C. Francis, L. Deanes, and C. Acey. 2022. "Climate change, environmental disasters, and health inequities: The underlying role of structural inequalities." *Current Environmental Health Reports* 9(1): 80–89. https://doi.org/10.1007/s40572-022-00336-w
111. Adams, Rachel I., Seema Bhangar, Karen C. Dannemiller, Jonathan A. Eisen, Noah Fierer, Jack A. Gilbert, and Jessica L. Green. 2015. "Ten questions concerning the microbiomes of buildings." *Building and Environment* 109: 224–234. https://doi.org/10.1016/j.buildenv.2016.09.001
112. Hoisington, Andrew J., Christopher E. Stamper, Katherine L. Bates, Maggie A. Stanislawski, Michael C. Flux, Teodor T. Postolache, Christopher A. Lowry, and Lisa A. Brenner. 2023. "Human microbiome transfer in the built environment differs based on occupants, objects, and buildings." *Scientific Reports* 13(1): 6446. https://doi.org/10.1038/s41598-023-33719-6
113. Taylor, Jonathon, Ka man Lai, Mike Davies, David Clifton, Ian Ridley, and Phillip Biddulph. 2011. "Flood management: Prediction of microbial contamination in large-scale floods in urban environments." *Environment International* 37 (5): 1019–1029. https://doi.org/10.1016/j.envint.2011.03.015
114. Peccia, Jordan, and Sarah E. Kwan. 2016. "Buildings, beneficial microbes, and health." *Trends in Microbiology* 24(8): 595–597. https://doi.org/10.1016/j.tim.2016.04.007
115. Dumon, Henri, Alain Palot, Carmel Charpin-Kadouch, Jacqueline Quéralt, Khrofia Lehtihet, Max Garans, and Denis Charpin. 2009. "Mold species identified in flooded dwellings." *Aerobiologia* 25(4): 341–344. https://doi.org/10.1007/s10453-009-9113-y
116. Riggs, Margaret A., Carol Y. Rao, Clive M. Brown, David Van Sickle, Kristin J. Cummings, Kevin H. Dunn, and James A. Deddens. 2008. "Resident cleanup activities, characteristics of flood-damaged homes and airborne microbial concentrations in New Orleans, Louisiana, October 2005." *Environmental Research* 106(3): 401–409. https://doi.org/10.1016/j.envres.2007.11.004
117. Cabrera, Pamela, Holly Samuelson, and Margaret Kurth. 2019. "Simulating mold risks under future climate conditions." In *Proceedings of Building Simulation 2019, Rome, Italy*. Vol. 16. Rome: International Building Performance Simulation Association.
118. Vecherin, Sergey, Matthew Joyner, Madison Smith, and Igor Linkov. 2024. "Risk assessment of mold growth across the US due to weather variations." *Building and Environment* 256: 111498. https://doi.org/10.1016/j.buildenv.2024.111498
119. American Society of Heating, Refrigerating and Air-Conditioning Engineers (ASHRAE). 2021. *Climatic Data for Building Design Standards*. Peachtree Corners, GA: American Society of Heating, Refrigerating and Air-Conditioning Engineers.
120. Kahn, David, William Chen, Yarrow Linden, Karalee A. Corbeil, Sarah Lowry, Ciara A. Higham, and Karla S. Mendez. 2024. "A microbial risk assessor's guide to valley fever (coccidioides spp.): Case study and review of risk factors." *Science of the Total Environment* 917: 170141. https://doi.org/10.1016/j.scitotenv.2024.170141; Gorris, Morgan E., Kathleen K. Treseder, Charles S. Zender, and James T. Randerson. 2019. "Expansion of coccidioidomycosis endemic regions in the United States in response to climate change." *Geohealth* 3(10): 308–327. https://doi.org/10.1029/2019GH000209
121. Huang, Jennifer Y., Benjamin Bristow, Shira Shafir, and Frank Sorvillo. 2012. "Coccidioidomycosis-associated deaths, United States, 1990–2008." *Emerging Infectious Diseases* 18(11): 1723. https://doi.org/10.3201%2Feid1811.120752
122. Fisk, William J., Wanyu R. Chan, and Alexandra L. Johnson. 2019. "Does dampness and mold in schools affect health? Results of a meta-analysis." *Indoor Air* 29(6): 895–902. https://doi.org/10.1111/ina.12588
123. Michaels, Robert A. 2017. "Environmental moisture, molds, and asthma—emerging fungal risks in the context of climate change." *Environmental Claims Journal* 29(3): 171–193. https://doi.org/10.1080/10406026.2017.1345521
124. Levin, Jay M., William R. Lewis, Heidi Hudson Raschke, Christina M. Phillips, Dennis C. Anderson, Sarah R. Burke, and Brian M. Collins. 2018. "Recent developments in property

insurance coverage litigation." *Tort Trial & Insurance Practice Law Journal* 53(2): 623–656. https://www.jstor.org/stable/27172766

125. Major, Jennifer L., and Gerald W. Boese. 2017. "Cross section of legislative approaches to reducing indoor dampness and mold." *Journal of Public Health Management and Practice* 23(4): 388–395. https://doi.org/10.1097/PHH.0000000000000491
126. Council of District of Columbia (CODC). 2019. Indoor Mold Remediation Enforcement Amendment Act of 2019. Washington, D.C.: Council of District of Columbia. Retrieved from http://lims.dccouncil.us/Download/41819/B23-0132-Introduction.pdf
127. Hauer, Mathew E., Jason M. Evans, and Deepak R. Mishra. 2016. "Millions projected to be at risk from sea-level rise in the continental United States." *Nature Climate Change* 6(7): 691–695. https://doi.org/10.1038/nclimate2961
128. Turney, Chris S. M., and Heidi Brown. 2007. "Catastrophic early Holocene sea level rise, human migration and the Neolithic transition in Europe." *Quaternary Science Reviews* 26(17–18): 2036–2041. https://doi.org/10.1016/j.quascirev.2007.07.003; Frumkin, Amos, Ofer Bar-Yosef, and Henry P. Schwarcz. 2011. "Possible paleohydrologic and paleoclimatic effects on hominin migration and occupation of the Levantine Middle Paleolithic." *Journal of Human Evolution* 60(4): 437–451. https://doi.org/10.1016/j.jhevol.2010.03.010; Kim, Hie Lim, Tanghua Li, Namrata Kalsi, Hung Tran The Nguyen, Timothy A. Shaw, Khai C. Ang, and Keith C. Cheng. 2023. "Prehistoric human migration between Sundaland and South Asia was driven by sea-level rise." *Communications Biology* 6(1): 150. https://doi.org/10.1038/s42003-023-04510-0
129. Anderson, David G., and Thaddeus G. Bissett. 2015. "The initial colonization of North America: Sea level change, shoreline movement, and great migrations." In *Mobility and Ancient Society in Asia and the Americas*, edited by Michael D. Frachetti and Robert N. Spengler III, 59–88. Cham: Springer International.
130. Thompson, Victor D., and John E. Worth. 2011. "Dwellers by the sea: Native American adaptations along the southern coasts of eastern North America." *Journal of Archaeological Research* (19): 51–101. https://doi.org/10.1007/s10814-010-9043-9
131. Thompson, Victor D., and John A. Turck. 2009. "Adaptive cycles of coastal hunter-gatherers." *American Antiquity* 74(2): 255–278. https://doi.org/10.1017/S0002731600048599
132. Hansen, James, Makiko Sato, Paul Hearty, Reto Ruedy, Maxwell Kelley, Valerie Masson-Delmotte, and Gary Russell. 2016. "Ice melt, sea level rise and superstorms: Evidence from paleoclimate data, climate modeling, and modern observations that 2°C global warming could be dangerous." *Atmospheric Chemistry and Physics* 16(6): 3761–3812. https://doi.org/10.5194/acp-16-3761-2016
133. Van Straalen, Fennie, Thomas Hartmann, and John Sheehan, eds. 2017. *Property Rights and Climate Change: Land Use Under Changing Environmental Conditions*. New York: Routledge.
134. Dahl, Kristina A., Erika Spanger-Siegfried, Astrid Caldas, and Shana Udvardy. 2017. "Effective inundation of continental United States communities with 21st century sea level rise." *Elementa: Science of the Anthropocene* 5: 37. https://doi.org/10.1525/elementa.234
135. Hoffman, John Steven, Dale L. Keyes, and James G. Titus. 1983. *Projecting Future Sea Level Rise: Methodology, Estimates to the Year 2100, and Research Needs*. Washington, D.C.: Environmental Protection Agency (EPA), Office of Policy and Resource Management. Retrieved from https://repository.library.noaa.gov/view/noaa/10299
136. Hoffman et al. 1983.
137. Frank, Thomas. 2019, April 11th. "After a $14-billion upgrade, New Orleans' levees are sinking." *Scientific American*. https://www.scientificamerican.com/article/after-a-14-billion-upgrade-new-orleans-levees-are-sinking/
138. Oppenheimer, M., B. C. Glavovic, J. Hinkel, R. van de Wal, A. K. Magnan, A. Abd-Elgawad, R. Cai, M. Cifuentes-Jara, R. M. DeConto, T. Ghosh, J. Hay, F. Isla, B. Marzeion, B. Meyssignac, and Z. Sebesvari. 2019. "Sea level rise and implications for low-lying islands, coasts and communities." In IPCC, *Special Report on the Ocean and Cryosphere in a Changing Climate*, edited by Pörtner, H.-O., D. C. Roberts, V. Masson-Delmotte, P. Zhai, M. Tignor, E. Poloczanska, K. Mintenbeck, A. Alegría, M. Nicolai, A. Okem, J. Petzold, B. Rama, N. M. Weyer. New York: Cambridge University Press. https://doi.org/10.1017/9781009157964.006; Beckley, B., X. Yang, N. P. Zelensky, S. A. Holmes, F. G. Lemoine, R. D. Ray, G. T. Mitchum, S. Desai, and S. T. Brown. 2021."Global mean sea level trend from integrated multi-mission ocean altimeters TOPEX/Poseidon, Jason-1, OSTM/Jason-2, and Jason-3 version 5.1." New York: NASA Goddard Space Flight Center. https://doi.org/10.5067/gmslm-tj151

139. May, C. L., M. S. Osler, H. F. Stockdon, P. L. Barnard, J. A. Callahan, R. C. Collini, C. M. Ferreira, J. Finzi Hart, E. E. Lentz, T. B. Mahoney, W. Sweet, D. Walker, and C. P. Weaver. 2023. "Ch. 9: Coastal effects." In Crimmins et al., *Fifth National Climate Assessment.*
140. May et al. 2023.
141. Coastal Protection and Restoration Authority. 2013. "A Changing Landscape." Baton Rouge: State of Louisiana. https://coastal.la.gov/whats-at-stake/a-changing-landscape/
142. Dangendorf, Sönke, Noah Hendricks, Qiang Sun, John Klinck, Tal Ezer, Thomas Frederikse, Francisco M. Calafat, Thomas Wahl, and Torbjörn E. Törnqvist. 2023. "Acceleration of US Southeast and Gulf coast sea-level rise amplified by internal climate variability." *Nature Communications* 14(1): 1–11. https://doi.org/10.1038/s41467-023-37649-9
143. Kayastha, Miraj B., Xinyu Ye, Chenfu Huang, and Pengfei Xue. 2022. "Future rise of the Great Lakes water levels under climate change." *Journal of Hydrology* 612: 128205. https://doi.org/10.1016/j.jhydrol.2022.128205
144. Walcott, R. I. 1972. "Late Quaternary vertical movements in eastern North America: Quantitative evidence of glacio-isostatic rebound." *Reviews of Geophysics* 10(4): 849–884. https://doi.org/10.1029/RG010i004p00849
145. Lee, Deborah H., and Charles F. Southam. 1994. "Effect and implications of differential isostatic rebound on Lake Superior's regulation limits." *Journal of Great Lakes Research* 20(2): 407–415. https://doi.org/10.1016/S0380-1330(94)71158-8
146. Sweet, W. V., B. D. Hamlington, R. E. Kopp, C. P. Weaver, P. L. Barnard, D. Bekaert, W. Brooks, M. Craghan, G. Dusek, T. Frederikse, G. Garner, A. S. Genz, J. P. Krasting, E. Larour, D. Marcy, J. J. Marra, J. Obeysekera, M. Osler, M. Pendleton, D. Roman, L. Schmied, W. Veatch, K. D. White, and C. Zuzak. 2022. *Global and Regional Sea Level Rise Scenarios for the United States: Updated Mean Projections and Extreme Water Level Probabilities Along U.S. Coastlines.* NOAA Technical Report NOS 01. Silver Spring, MD: National Oceanic and Atmospheric Administration, National Ocean Service. https://oceanservice.noaa.gov/hazards/sealevelrise/noaa-nos-techrpt01-global-regional-SLR-scenarios-US.pdf
147. National Oceanographic and Atmospheric Administration (NOAA). 2023. "Climate change: Ocean heat content." Silver Spring, MD: National Oceanographic and Atmospheric Administration. https://www.climate.gov/news-features/understanding-climate/climate-change-ocean-heat-content
148. See note 146.
149. Naughten, Kaitlin A., Paul R. Holland, and Jan De Rydt. 2023. "Unavoidable future increase in West Antarctic ice-shelf melting over the twenty-first century." *Nature Climate Change* 13(11): 1222–1228. https://doi.org/10.1038/s41558-023-01818-x
150. Gornitz, Vivien M., Michael Oppenheimer, Robert Kopp, Philip Orton, Maya Buchanan, Ning Lin, Radley Horton, and Daniel A. Bader. "New York City Panel on Climate Change 2019 report chapter 3: Sea level rise." *Annals of the New York Academy of Sciences.* https://doi.org/10.1111/nyas.14006
151. Folkers, Andreas. 2021. "Fossil modernity: The materiality of acceleration, slow violence, and ecological futures." *Time & Society* 30(1): 223–246. https://doi.org/10.1177/0961463X20987965
152. Derluyn, Hannelore, Peter Vontobel, David Mannes, Dominique Derome, Eberhard Lehmann, and Jan Carmeliet. 2019. "Saline water evaporation and crystallization-induced deformations in building stone: Insights from high-resolution neutron radiography." *Transport in Porous Media* 128: 895–913. https://doi.org/10.1007/s11242-018-1151-x
153. Habel, Shellie, Charles H. Fletcher, Matthew M. Barbee, and Kyrstin L. Fornace. 2024. "Hidden threat: The influence of sea-level rise on coastal groundwater and the convergence of impacts on municipal infrastructure." *Annual Review of Marine Science* 16: 81–103. https://doi.org/10.1146/annurev-marine-020923-120737
154. McKenzie, Tristan, Shellie Habel, and Henrietta Dulai. 2021. "Sea-level rise drives wastewater leakage to coastal waters and storm drains." *Limnology and Oceanography Letters* 6(3): 154–163. https://doi.org/10.1002/lol2.10186
155. Author's laboratory calculations; McBride, Matt, 2023, November 8th. "Tapped out: New Orleans drinking water testing procedures don't follow gov't regulations." *Louisiana Illuminator.* https://lailluminator.com/2023/11/08/new-orleans-water/
156. Heiss, James W., Bryce Mase, and Chengji Shen. 2022. "Effects of future increases in tidal flooding on salinity and groundwater dynamics in coastal aquifers." *Water Resources Research* 58(12): e2022WR033195. https://doi.org/10.1029/2022WR033195

157. Hill, Kristina, Daniella Hirschfeld, C. Lindquist, Forest Cook, and Scott Warner. 2023. "Rising coastal groundwater as a result of sea-level rise will influence contaminated coastal sites and underground infrastructure." *Earth's Future* 11(9): e2023EF003825. https://doi.org/10.1029/2023EF003825
158. Flavelle, Christopher. 2018, August 29th. "Miami will be underwater soon: Its drinking water could go first." *Bloomberg News*. https://www.bloomberg.com/news/features/2018-08-29/miami-s-other-water-problem
159. Flanagan, Kyle, Barnali Dixon, Tess Rivenbark, and Dale Griffin. 2020. "An integrative GIS approach to analyzing the impacts of septic systems on the coast of Florida, USA." *Physical Geography* 41(5): 407–432. https://doi.org/10.1080/02723646.2019.1671297
160. Environmental Protection Agency (EPA). 2023. *Small Wastewater Systems Research*. Washington, D.C.: Environmental Protection Agency.
161. Dennis, Brady, Kevin Crowe and John Muyskens. 2024, May 22nd. "A hidden threat: Fast-rising seas could swamp septic systems in parts of the South." *The Washington Post*. https://www.washingtonpost.com/climate-environment/interactive/2024/septic-tanks-rising-waters-environment-health/
162. Brzezinski, Adam, Amol Purandare, and Samit Patel. 2024. "*Vibrio cholerae* in Florida: An unexpected presentation." *Clinical Pediatrics* 63(11): 1601–1603. https://doi.org/10.1177/00099228241228108
163. Rocklöv, Joacim, and Robert Dubrow. "Climate change: An enduring challenge for vector-borne disease prevention and control." *Nature Immunology* 21(5): 479–483. https://doi.org/10.1038/s41590-020-0648-y
164. Wilke, André B. B., Barbara F. Wisinski, Giovanni Benelli, Chalmers Vasquez, John-Paul Mutebi, William D. Petrie, and John C. Beier. 2021. "Local conditions favor dengue transmission in the contiguous United States." *Entomologia Generalis* 41(5). 523. https://doi.org/10.1127/entomologia/2021/1202; Wilke, André B. B., Chalmers Vasquez, Johana Medina, Isik Unlu, John C. Beier, and Marco Ajelli. 2024. "Presence and abundance of malaria vector species in Miami-Dade County, Florida." *Malaria Journal* 23(1): 24. https://doi.org/10.1186/s12936-024-04847-9
165. Mora, Camilo, Tristan McKenzie, Isabella M. Gaw, Jacqueline M. Dean, Hannah von Hammerstein, Tabatha A. Knudson, and Renee O. Setter. 2022. "Over half of known human pathogenic diseases can be aggravated by climate change." *Nature Climate Change* 12(9): 869–875. https://doi.org/10.1038/s41558-022-01426-1
166. Martello, Michael V., Andrew J. Whittle, Jesse M. Keenan, and Frederick P. Salvucci. 2021. "Evaluation of climate change resilience for Boston's rail rapid transit network." *Transportation Research Part D: Transport and Environment* 97: 102908. https://doi.org/10.1016/j.trd.2021.102908
167. Coffel, Ethan D., Terence R. Thompson, and Radley M. Horton. 2017. "The impacts of rising temperatures on aircraft takeoff performance." *Climatic Change* 144: 381–388. https://doi.org/10.1007/s10584-017-2018-9
168. Buchanan, Maya K., Scott Kulp, Lara Cushing, Rachel Morello-Frosch, Todd Nedwick, and Benjamin Strauss. 2020. "Sea level rise and coastal flooding threaten affordable housing." *Environmental Research Letters* 15(12): 124020. https://doi.org/10.1088/1748-9326/abb266
169. NYU Furman Center. 2017. "Housing in the U.S. floodplains." Data brief. New York: New York University. Retrieved from https://furmancenter.org/files/NYUFurmanCenter_HousingInTheFloodplain_May2017.pdf
170. Crawford, Susan. 2023. *Charleston: Race, Water, and the Coming Storm*. New York: Simon and Schuster.
171. Crawford 2023.
172. Handwerger, Leah R., Margaret M. Sugg, and Jennifer D. Runkle. 2021. "Present and future sea level rise at the intersection of race and poverty in the Carolinas: A geospatial analysis." *The Journal of Climate Change and Health* 3: 100028. https://doi.org/10.1016/j.joclim.2021.100028
173. Anderson, David G., Thaddeus G. Bissett, Stephen J. Yerka, Joshua J. Wells, Eric C. Kansa, Sarah W. Kansa, Kelsey Noack Myers, R. Carl DeMuth, and Devin A. White. 2017. "Sea-level rise and archaeological site destruction: An example from the southeastern United States using DINAA (Digital index of North American archaeology)." *PLOS One* 12(11): e0188142. https://doi.org/10.1371/journal.pone.0188142

174. Seekamp, Erin, Sandra Fatorić, and Allie McCreary. 2020. “Historic preservation priorities for climate adaptation.” *Ocean & Coastal Management* 191: 105180. https://doi.org/10.1016/j.ocecoaman.2020.105180
175. McCoy, Sara Brent. 2020. “Climate as provocation of preservation standards and procedure in historic districts of the floodprone US: Lessons from Palm View, Miami Beach.” Master’s thesis. Cambridge, MA., Massachusetts Institute of Technology. https://dspace.mit.edu/handle/1721.1/128970
176. Hauer, Mathew E., Dean Hardy, Scott A. Kulp, Valerie Mueller, David J. Wrathall, and Peter U. Clark. 2021. “Assessing population exposure to coastal flooding due to sea level rise.” *Nature Communications* 12(1): 6900. https://doi.org/10.1038/s43017-019-0002-9
177. Hauer et al. 2021.
178. See note 127.
179. Logan, Tom, Mitchell J. Anderson, and Allison C. Reilly. 2023. “Risk of isolation increases the expected burden from sea-level rise.” *Nature Climate Change* 13(4): 397–402. https://doi.org/10.1038/s41558-023-01642-3
180. Sheldon, Tamara L., and Crystal Zhan. 2022. “The impact of hurricanes and floods on domestic migration.” *Journal of Environmental Economics and Management* 115: 102726. https://doi.org/10.1016/j.jeem.2022.102726
181. Smith, Stanley K. 1996. “Demography of disaster: Population estimates after Hurricane Andrew.” *Population Research and Policy Review* 15: 459–477. https://doi.org/10.1007/BF00125865
182. Balaguru, Karthik, Wenwei Xu, Chuan-Chieh Chang, L. Ruby Leung, David R. Judi, Samson M. Hagos, Michael F. Wehner, James P. Kossin, and Mingfang Ting. 2023. “Increased US coastal hurricane risk under climate change.” *Science Advances* 9(14): eadf0259. https://doi.org/10.1126/sciadv.adf0259
183. Emanuel, Kerry. 2005. “Increasing destructiveness of tropical cyclones over the past 30 years.” *Nature* 436(7051): 686–688. https://doi.org/10.1038/nature03906; Saunders, Mark A., and Adam S. Lea. 2008. “Large contribution of sea surface warming to recent increase in Atlantic hurricane activity.” *Nature* 451(7178): 557–560. https://doi.org/10.1038/nature06422
184. El Niño-Southern Oscillation (ENSO) Atlantic Multidecadal Oscillation (AMO)
185. Méndez-Tejeda, Rafael, and José J. Hernández-Ayala. 2023. “Links between climate change and hurricanes in the North Atlantic.” *PLoS Climate* 2(4): e0000186. https://doi.org/10.1371/journal.pclm.0000186
186. National Hurricane Center. 2024. “Storm surge overview.” Miami: National Oceanographic and Atmospheric Administration. https://www.nhc.noaa.gov/surge/#THREAT
187. Masoomi, Hassan, John W. van de Lindt, Mohammad R. Ameri, Trung Q. Do, and Bret M. Webb. 2019. “Combined wind-wave-surge hurricane-induced damage prediction for buildings.” *Journal of Structural Engineering* 145(11): 04018227. https://doi.org/10.1061/(ASCE)ST.1943-541X.0002241
188. Congressional Budget Office. 2019. “Expected costs of damages from hurricanes winds and storm-related flooding.” Washington, D.C.: U.S. Congress. https://www.cbo.gov/publication/55019#section0
189. Simmons, Kevin M., Jeffrey Czajkowski, and James M. Done. 2018. “Economic effectiveness of implementing a statewide building code: The case of Florida.” *Land Economics* 94(2): 155–174. https://doi.org/10.3368/le.94.2.155
190. Simmons et al. 2018.
191. State of Florida Department of Business and Professional Regulation, Florida Building Commission and University of Florida, GeoPlan Center. 2023. *Draft Final Report: Update and Development of Wind Speed Line Maps for the Florida Building Code, 8th Edition*. Tallahassee: Florida Building Commission. https://www.floridabuilding.org/fbc/commission/FBC_0623/Structural_TAC/Draft_Final_Report_Wind_Speed_UF_20230608%20-%20Copy.pdf
192. Ollila, Antero. 2023. “Radiative forcing and climate sensitivity of carbon dioxide (CO_2) fine-tuned with CERES data.” *Current Journal of Applied Science and Technology* 42(46): 111–133. https://doi.org/10.9734/cjast/2023/v42i464300
193. Intergovernmental Panel on Climate Change (IPCC). 2023. “Summary for Policymakers.” In *Climate Change 2023: Synthesis Report. Contribution of Working Groups I, II and III to the Sixth Assessment Report of the Intergovernmental Panel on Climate Change*, edited by Core

Writing Team, H. Lee and J. Romero. Geneva: IPCC. https://doi.org/10.59327/IPCC/AR6-9789291691647.001
194. National Centers for Environmental Information 2024.
195. National Oceanographic and Atmospheric Administration (NOAA). 2025. "2024 was the world's warmest year on record." Washington, D.C.: National Oceanographic and Atmospheric Administration (NOAA). https://www.noaa.gov/news/2024-was-worlds-warmest-year-on-record
196. Yoshioka, Masaru, Daniel P. Grosvenor, Ben B. B. Booth, Colin P. Morice, and Kenneth S. Carslaw. 2024. "Warming effects of reduced sulfur emissions from shipping." *EGUsphere Preprint*, 1–19. https://doi.org/10.5194/egusphere-2024-1428
197. Yoshioka et al. 2024.
198. Yuan, Tianle, Hua Song, Lazaros Oreopoulos, Robert Wood, Huisheng Bian, Katherine Breen, and Mian Chin. 2024. "Abrupt reduction in shipping emission as an inadvertent geoengineering termination shock produces substantial radiative warming." *Communications Earth & Environment* 5(1): 281. https://doi.org/10.1038/s43247-024-01442-3
199. Marvel et al. 2023.
200. Marvel et al. 2023.
201. Kuczyński, Tadeusz, Anna Staszczuk, Marta Gortych, and Roman Stryjski. 2021. "Effect of thermal mass, night ventilation and window shading on summer thermal comfort of buildings in a temperate climate." *Building and Environment* 204: 108126.
202. Minor, Kelton, Andreas Bjerre-Nielsen, Sigga Svala Jonasdottir, Sune Lehmann, and Nick Obradovich. 2022. "Rising temperatures erode human sleep globally." *One Earth* 5(5): 534–549. https://doi.org/10.1016/j.oneear.2022.04.008
203. He, Cheng, Ho Kim, Masahiro Hashizume, Whanhee Lee, Yasushi Honda, Satbyul Estella Kim, and Patrick L. Kinney. 2022. "The effects of night-time warming on mortality burden under future climate change scenarios: a modelling study." *The Lancet Planetary Health* 6(8): e648–e657. https://doi.org/10.1016/S2542-5196(22)00139-5
204. Su, Bin, Renata Jadresin Milic, Peter McPherson, and Lian Wu. 2022. "Thermal performance of school buildings: Impacts beyond thermal comfort." *International Journal of Environmental Research and Public Health* 19(10): 5811. https://doi.org/10.3390/ijerph19105811
205. Eijkelenboom, Anne Marie, and Philomena M. Bluyssen. 2022. "Comfort and health of patients and staff, related to the physical environment of different departments in hospitals: A literature review." *Intelligent Buildings International* 14(1): 95–113. https://doi.org/10.1080/17508975.2019.1613218
206. Park, R. Jisung, Joshua Goodman, Michael Hurwitz, and Jonathan Smith. 2020. "Heat and learning." *American Economic Journal: Economic Policy* 12(2): 306–39. https://doi.org/10.1257/pol.20180612
207. Government Accountability Office (GAO). 2020, June. *K–12 Education: School Districts Frequently Identified Multiple Building Systems Needing Updates or Replacement.* Washington, D.C.: U.S. Congress. https://www.gao.gov/assets/710/707517.pdf
208. Author's calculations.
209. U.S. Global Change Research Program (USGCRP). 2024. "Heating and cooling degree days." Washington, D.C.: Office of Science and Technology Policy, Executive Office of the President of the United States. Retrieved from https://www.globalchange.gov/indicators/heating-and-cooling-degree-days
210. U.S. Energy Information Administration (EIA). "Nearly 90% of U.S. households used air conditioning in 2020." Washington, D.C.: U.S. Department of Energy. Retrieved from https://www.eia.gov/todayinenergy/detail.php?id=52558
211. Williams, Augusta, Larissa McDonogh-Wong, and John D. Spengler. 2020. "The influence of extreme heat on police and fire department services in 23 US cities." *GeoHealth* 4(11): e2020GH000282. https://doi.org/10.1029/2020GH000282
212. Abel, David W., Tracey Holloway, Monica Harkey, Paul Meier, Doug Ahl, Vijay S. Limaye, and Jonathan A. Patz. 2018. "Air-quality-related health impacts from climate change and from adaptation of cooling demand for buildings in the eastern United States: An interdisciplinary modeling study." *PLoS Medicine* 15(7): e1002599. https://doi.org/10.1371/journal.pmed.1002599
213. Stone, Brian, Jr., Evan Mallen, Mayuri Rajput, Carina J. Gronlund, Ashley M. Broadbent, E. Scott Krayenhoff, Godfried Augenbroe, Marie S. O'Neill, and Matei Georgescu. 2021.

"Compound climate and infrastructure events: How electrical grid failure alters heat wave risk." *Environmental Science & Technology* 55(10): 6957–6964. https://doi.org/10.1021/acs.est.1c00024

214. Baniassadi, Amir, David J. Sailor, E. Scott Krayenhoff, Ashley M. Broadbent, and Matei Georgescu. 2019. "Passive survivability of buildings under changing urban climates across eight US cities." *Environmental Research Letters* 14 (7): 074028. https://doi.org/10.1088/1748-9326/ab28ba
215. See note 72; U.S. Global Research Program (USGCRP). 2024. "USGCRP indicators platform: Heat waves." Washington, D.C.: Office of Science and Technology Policy, Executive Office of the President of the United States. https://www.globalchange.gov/indicators/heat-waves
216. Wang, Tianni, Zhuohua Qu, Zaili Yang, Timothy Nichol, Geoff Clarke, and Ying-En Ge. 2020. "Climate change research on transportation systems: Climate risks, adaptation and planning." *Transportation Research Part D: Transport and Environment* 88: 102553. https://doi.org/10.1016/j.trd.2020.102553
217. Centers for Disease Control and Prevention (CDC). 2024. "Extreme heat and your health." Atlanta: Centers for Disease Control and Prevention. Retrieved from https://www.cdc.gov/extreme-heat/about/index.html
218. Marvel et al. 2023.
219. Marvel et al. 2023.
220. Ray, Bhaswati, and S. Rajib, eds. 2019. *Urban Drought*. Singapore: Springer International.
221. Obringer, Renee, Roshanak Nateghi, Jessica Knee, Kaveh Madani, and Rohini Kumar. 2024. "Urban water and electricity demand data for understanding climate change impacts on the water-energy nexus." *Scientific Data* 11(1): 108. https://doi.org/10.1038/s41597-024-02930-z
222. Gurmu, Argaw Tarekegn, Adam Krezel, and Muhammad Nateque Mahmood. 2023. "Analysis of the causes of defects in ground floor systems of residential buildings." *International Journal of Construction Management* 23(2): 268–275. https://doi.org/10.1080/15623599.2020.1860636
223. Flavelle, Christopher, and Jack Healy. 2023, June 1st. "Arizona limits construction around Phoenix as its water supply dwindles." *The New York Times*. https://www.nytimes.com/2023/06/01/climate/arizona-phoenix-permits-housing-water.html
224. Wirz, Matt. 2024, August 25th. "This Texas city is too hot, short on water—and booming." *The Wall Street Journal*. https://www.wsj.com/us-news/climate-environment/kyle-texas-city-growth-heat-water-6660dc42
225. Wirz 2024.
226. Monroe, Rachel. 2022, June 29th. "The water wars come to the suburbs." *The New Yorker*. https://www.newyorker.com/news/letter-from-the-southwest/the-water-wars-come-to-the-suburbs
227. Duthu, Ray C., and Thomas H. Bradley. 2017. "A road damage and life-cycle greenhouse gas comparison of trucking and pipeline water delivery systems for hydraulically fractured oil and gas field development in Colorado." *PLOS One* 12(7): e0180587. https://doi.org/10.1371/journal.pone.0180587
228. Goodwin, Marissa J., Harold S. J. Zald, Malcolm P. North, and Matthew D. Hurteau. 2021. "Climate-driven tree mortality and fuel aridity increase wildfire's potential heat flux." *Geophysical Research Letters* 48(24): e2021GL094954. https://doi.org/10.1029/2021GL094954
229. Dale, Virginia H., Linda A. Joyce, Steve McNulty, Ronald P. Neilson, Matthew P. Ayres, Michael D. Flannigan, and Paul J. Hanson. 2001. "Climate change and forest disturbances: Climate change can affect forests by altering the frequency, intensity, duration, and timing of fire, drought, introduced species, insect and pathogen outbreaks, hurricanes, windstorms, ice storms, or landslides." *BioScience* 51(9): 723–734. https://doi.org/10.1641/0006-3568(2001)051[0723:CCAFD]2.0.CO;2
230. Stephens, Scott L., A. LeRoy Westerling, Matthew D. Hurteau, M. Zachariah Peery, Courtney A. Schultz, and Sally Thompson. 2020. "Fire and climate change: Conserving seasonally dry forests is still possible." *Frontiers in Ecology and the Environment* 18(6): 354–360. https://doi.org/10.1002/fee.2218
231. See note 72.
232. Carlson, Amanda R., David P. Helmers, Todd J. Hawbaker, Miranda H. Mockrin, and Volker C. Radeloff. 2022. "The wildland-urban interface in the United States based on 125 million building locations." *Ecological Applications* 32(5): e2597. https://doi.org/10.1002/eap.2597

233. Mockrin, Miranda H., David Helmers, Sebastian Martinuzzi, Todd J. Hawbaker, and Volker C. Radeloff. 2022. "Growth of the wildland-urban interface within and around US National Forests and Grasslands, 1990–2010." *Landscape and Urban Planning* 218: 104283. https://doi.org/10.1016/j.landurbplan.2021.104283
234. Burke, Marshall, Anne Driscoll, Sam Heft-Neal, Jiani Xue, Jennifer Burney, and Michael Wara. 2021. "The changing risk and burden of wildfire in the United States." *Proceedings of the National Academy of Sciences* 118(2): e2011048118. https://doi.org/10.1073/pnas.2011048118
235. Caggiano, Michael D., Todd J. Hawbaker, Benjamin M. Gannon, and Chad M. Hoffman. 2020. "Building loss in WUI disasters: Evaluating the core components of the wildland-urban interface definition." *Fire* 3(4): 73. https://doi.org/10.3390/fire3040073
236. Radeloff, Volker C., Miranda H. Mockrin, David Helmers, Amanda Carlson, Todd J. Hawbaker, Sebastian Martinuzzi, Franz Schug, Patricia M. Alexandre, H. Anu Kramer, and Anna M. Pidgeon. 2023. "Rising wildfire risk to houses in the United States, especially in grasslands and shrublands." *Science* 382(6671): 702–707. https://doi.org/10.1126/science.ade9223
237. Centers for Disease Control and Prevention (CDC). 2024. "Climate and health: Wildfires." Atlanta: Centers for Disease Control and Prevention. https://www.cdc.gov/climate-health/php/effects/wildfires.html
238. National Institute for Occupational Safety and Health (NIOSH). 2023. "Fighting wildfires." Atlanta: Centers for Disease Control and Prevention. https://www.cdc.gov/niosh/firefighters/about/wildfires.html
239. Baylis, Patrick W., and Judson Boomhower. 2022. "Mandated vs. voluntary adaptation to natural disasters: The case of US wildfires." Working Paper. No. 29621. Cambridge, MA: National Bureau of Economic Research. https://doi.org/10.3386/w29621
240. Higuera, Philip E., Maxwell C. Cook, Jennifer K. Balch, E. Natasha Stavros, Adam L. Mahood, and Lise A. St. Denis. 2023. "Shifting social-ecological fire regimes explain increasing structure loss from Western wildfires." *PNAS Nexus* 2(3): pgad005. https://doi.org/10.1093/pnasnexus/pgad005
241. Fernandez-Marcos, and Maria Luisa. 2022. "Potentially toxic substances and associated risks in soils affected by wildfires: A review." *Toxics* 10(1): 31. https://doi.org/10.3390/toxics10010031
242. Reisen, Fabienne, Mahendra Bhujel, and Justin Leonard. 2014. "Particle and volatile organic emissions from the combustion of a range of building and furnishing materials using a cone calorimeter." *Fire Safety Journal* 69: 76–88. https://doi.org/10.1016/j.firesaf.2014.08.008; Metz, Amy J., Erica C. Fischer, and Brad P. Wham. 2023. "Behavior of service lateral pipes during wildfires: Testing methodologies and impact on drinking water contamination." *Environmental Science and Technology: Water* 3(2): 275–286. https://doi.org/10.1021/acsestwater.2c00248
243. Touma, Danielle, Samantha Stevenson, Daniel L. Swain, Deepti Singh, Dmitri A. Kalashnikov, and Xingying Huang. 2022. "Climate change increases risk of extreme rainfall following wildfire in the western United States." *Science Advances* 8(13): eabm0320. https://doi.org/10.1126/sciadv.abm0320
244. May, Nathaniel W., Clara Dixon, and Daniel A. Jaffe. 2021. "Impact of wildfire smoke events on indoor air quality and evaluation of a low-cost filtration method." *Aerosol and Air Quality Research* 21(7): 210046. https://doi.org/10.4209/aaqr.210046
245. Saunders, Rolando O., and Darryn W. Waugh. 2015. "Variability and potential sources of summer PM2. 5 in the Northeastern United States." *Atmospheric Environment* 117: 259–270. https://doi.org/10.1016/j.atmosenv.2015.07.007; Burke, Marshall, Marissa L. Childs, Brandon de la Cuesta, Minghao Qiu, Jessica Li, Carlos F. Gould, Sam Heft-Neal, and Michael Wara. 2024. "The contribution of wildfire to PM2.5 trends in the USA." *Nature* 622(7984): 761–766. https://doi.org/10.1038/s41586-023-06522-6
246. McArdle, Cristin E., Tia Dowling, Kelly Carey, Jourdan DeVies, Dylan Johs, Abigail L. Gates, Zachary Stein, Katharina L. van Santen, Lakshimi Radhakrishran, Aaron Kite-Powel, Karl Soetebier, Jason D. Sacks, Kanta Sircar, Kathleen P. Hartnett, and Maria C. Mirabelli. 2023. "Asthma-associated emergency department visits during the Canadian wildfire smoke episodes—United States, April–August 2023." *CDC: Morbidity and Mortality Weekly Report* 72(34): 926–932. http://dx.doi.org/10.15585/mmwr.mm7234a5
247. Jiang, P., Y. Li, M. K. Tong, S. Ha, E. Gaw, J. Nie, P. Mendola, and M. Wang. 2024. "Wildfire particulate exposure and risks of preterm birth and low birth weight in the Southwestern United States." *Public Health* 230: 81–88. https://doi.org/10.1016/j.puhe.2024.02.016

248. Molitor, David, Jamie T. Mullins, and Corey White. 2023. "Air pollution and suicide in rural and urban America: Evidence from wildfire smoke." *Proceedings of the National Academy of Sciences* 120(38): e2221621120. https://doi.org/10.1073/pnas.2221621120
249. Van Leeuwen, Cornelis, Giovanni Sgubin, Benjamin Bois, Nathalie Ollat, Didier Swingedouw, Sébastien Zito, and Gregory A. Gambetta. 2024. "Climate change impacts and adaptations of wine production." *Nature Reviews Earth & Environment* 5(4): 258–275. https://doi.org/10.1038/s43017-024-00521-5

Chapter 3

1. Climate change has occurred throughout the course of earth's existence. References to climate change herein are a direct reference to anthropogenic climate change, unless otherwise noted.
2. Umina, Paul Anthony, Andrew R. Weeks, Michael R. Kearney, Stephen William McKechnie, and Ary A. Hoffmann. 2005. "A rapid shift in a classic clinal pattern in *Drosophila* reflecting climate change." *Science* 308(5722): 691–693. https://doi.org/10.1126/science.1109523
3. Emerson, Kevin J., Clayton R. Merz, Julian M. Catchen, Paul A. Hohenlohe, William A. Cresko, William E. Bradshaw, and Christina M. Holzapfel. 2010. "Resolving postglacial phylogeography using high-throughput sequencing." *Proceedings of the National Academy of Sciences* 107(37): 16196–16200. https://doi.org/10.1073/pnas.1006538107; National Science Foundation. 2010, August 10th. "Genetic structure of first animal to show evolutionary response to climate change determined." Phys.org. https://phys.org/news/2010-08-genetic-animal-evolutionary-response-climate.html
4. Bradshaw, William E., Kevin J. Emerson, Julian M. Catchen, William A. Cresko, and Christina M. Holzapfel. 2012. "Footprints in time: Comparative quantitative trait loci mapping of the pitcher-plant mosquito, *Wyeomyia smithii*." *Proceedings of the Royal Society B: Biological Sciences* 279(1747): 4551–4558. https://doi.org/10.1007/s00359-023-01643-9
5. Oostra, Vicencio, Marjo Saastamoinen, Bas J. Zwaan, and Christopher W. Wheat. 2018. "Strong phenotypic plasticity limits potential for evolutionary responses to climate change." *Nature Communications* 9(1):1005. https://doi.org/10.1038/s41467-018-03384-9
6. Kelly, Morgan. 2019. "Adaptation to climate change through genetic accommodation and assimilation of plastic phenotypes." *Philosophical Transactions of the Royal Society B* 374(1768): 20180176. https://doi.org/10.1098/rstb.2018.0176
7. Waller, Natalie L., Ian C. Gynther, Alastair B. Freeman, Tyrone H. Lavery, and Luke K.-P. Leung. 2017. "The Bramble Cay melomys *Melomys rubicola* (Rodentia: Muridae): A first mammalian extinction caused by human-induced climate change?" *Wildlife Research* 44(1): 9–21. https://doi.org/10.1071/WR16157
8. Román-Palacios, Cristian, and John J. Wiens. 2020. "Recent responses to climate change reveal the drivers of species extinction and survival." *Proceedings of the National Academy of Sciences* 117(8): 4211–4217. https://doi.org/10.1073/pnas.1913007117
9. Parmesan, Camille. 2006. "Ecological and evolutionary responses to recent climate change." *Annual Review of Ecology, Evolution, and Systematics* 37(1): 637–669. https://doi.org/10.1146/annurev.ecolsys.37.091305.110100
10. Coope, G. Russell. 1994. "The response of insect faunas to glacial-interglacial climatic fluctuations." *Philosophical Transactions of the Royal Society of London. Series B: Biological Sciences* 344(1307): 19–26. https://doi.org/10.1098/rstb.1994.0046
11. Naylor, Angus, James Ford, Tristan Pearce, and James Van Alstine. 2020. "Conceptualizing climate vulnerability in complex adaptive systems." *One Earth* 2(5): 444–454. https://doi.org/10.1016/j.oneear.2020.04.011
12. Jackson, Michelle C., Samraat Pawar, and Guy Woodward. 2021. "The temporal dynamics of multiple stressor effects: From individuals to ecosystems." *Trends in Ecology & Evolution* 36(5): 402–410. https://doi.org/10.1016/j.tree.2021.01.005
13. Burkhard, Benjamin, Brian D. Fath, and Felix Müller. 2011. "Adapting the adaptive cycle: Hypotheses on the development of ecosystem properties and services." *Ecological Modelling* 222(11): 2878–2890. https://doi.org/10.1016/j.ecolmodel.2011.05.016
14. Gunderson, Lance H., and Crawford Stanley Holling. 2002. *Panarchy: Understanding Transformations in Human and Natural Systems*. Washington, D.C.: Island Press.
15. Sundstrom, Shana M., and Craig R. Allen. 2019. "The adaptive cycle: More than a metaphor." *Ecological Complexity* 39: 100767. https://doi.org/10.1016/j.ecocom.2019.100767
16. Sundstrom and Allen 2019.

17. Holling, Crawford S. 1973. "Resilience and stability of ecological systems." *Annual Review of Ecology and Systematics* 4: 1–23. https://doi.org/10.1017/9781009177856.038
18. Holling 1973.
19. Gunderson, Lance H. 2000. "Ecological resilience—in theory and application." *Annual Review of Ecology and Systematics* 31(1): 425–439. https://doi.org/10.1146/annurev.ecolsys.31.1.425
20. Folke, Carl. 2006. "Resilience: The emergence of a perspective for social-ecological systems analyses." *Global Environmental Change* 16(3): 253–267. https://doi.org/10.1016/j.gloenvcha.2006.04.002
21. Keenan, Jesse M. 2018. "Types and forms of resilience in local planning in the U.S.: Who does what?" *Environmental Science & Policy* 88: 116–123. https://doi.org/10.1016/j.envsci.2018.06.015
22. Folke 2006.
23. Cumming, Graeme S., Grenville Barnes, Stephen Perz, Marianne Schmink, Kathryn E. Sieving, Jane Southworth, Michael Binford, Robert D. Holt, Claudia Stickler, and Tracy Van Holt. 2005. "An exploratory framework for the empirical measurement of resilience." *Ecosystems* 8: 975–987. https://doi.org/10.1007/s10021-005-0129-z
24. Andrachuk, Mark, and Derek Armitage. 2015. "Understanding social-ecological change and transformation through community perceptions of system identity." *Ecology and Society* 20(4): 26. http://dx.doi.org/10.5751/ES-07759-200426
25. This book liberally uses the term "system" to include a variety of objects, organizations, processes, and relationships. Non-systems dynamics and conditions may also be at work.
26. Olsson, Lennart, Anne Jerneck, Henrik Thoren, Johannes Persson, and David O'Byrne. 2015. "Why resilience is unappealing to social science: Theoretical and empirical investigations of the scientific use of resilience." *Science Advances* 1(4): e1400217. https://doi.org/10.1126/sciadv.1400217
27. Andrachuk and Armitage 2015.
28. Delettre, Olivier. 2021. "Identity of ecological systems and the meaning of resilience." *Journal of Ecology* 109 (9): 3147–3156. https://doi.org/10.1111/1365-2745.13655
29. Pelling, Mark, Karen O'Brien, and David Matyas. 2015. "Adaptation and transformation." *Climatic Change* 133: 113–127. https://doi.org/10.1007/s10584-014-1303-0
30. IPCC 2022.
31. Kates, Robert W., William R. Travis, and Thomas J. Wilbanks. 2012. "Transformational adaptation when incremental adaptations to climate change are insufficient." *Proceedings of the National Academy of Sciences* 109(19): 7156–7161. https://doi.org/10.1073/pnas.1115521109
32. Pelling, O'Brien, and Matyas 2015.
33. Patterson, James, Douwe L. de Voogt, and Rodolfo Sapiains. 2019. "Beyond inputs and outputs: Process-oriented explanation of institutional change in climate adaptation governance." *Environmental Policy and Governance* 29(5): 360–375. https://doi.org/10.1002/eet.1865
34. IPCC 2022.
35. Juhola, Sirkku, Erik Glaas, Björn-Ola Linnér, and Tina-Simone Neset. 2016. "Redefining maladaptation." *Environmental Science & Policy* 55: 135–140. https://doi.org/10.1016/j.envsci.2015.09.014
36. Dow, Kirstin, Frans Berkhout, Benjamin L. Preston, Richard J. T. Klein, Guy Midgley, and M. Rebecca Shaw. 2013. "Limits to adaptation." *Nature Climate Change* 3(4): 305–307. https://doi.org/10.1038/nclimate1847
37. Rickards, Lauren. 2013. "Transformation is adaptation." *Nature Climate Change* 3(8): 690–690. https://doi.org/10.1038/nclimate1933
38. IPCC 2022.
39. Morrison, Tiffany H., W. Neil Adger, Katrina Brown, Maria Carmen Lemos, Dave Huitema, and Terry P. Hughes. 2017. "Mitigation and adaptation in polycentric systems: Sources of power in the pursuit of collective goals." *Wiley Interdisciplinary Reviews: Climate Change* 8(5): e479. https://doi.org/10.1002/wcc.479
40. Adger, W. Neil, Nigel W. Arnell, and Emma L. Tompkins. 2005. "Successful adaptation to climate change across scales." *Global Environmental Change* 15(2): 77–86. https://doi.org/10.1016/j.gloenvcha.2004.12.005
41. Smit, Barry, Ian Burton, Richard J. T. Klein, and Johanna Wandel. 2000. "An anatomy of adaptation to climate change and variability." *Climatic Change* 45(1): 223–251. https://doi.org/10.1007/978-94-017-3010-5_12

42. Barnett, Jon, Louisa S. Evans, Catherine Gross, Anthony S. Kiem, Richard T. Kingsford, Jean P. Palutikof, Catherine M. Pickering, and Scott G. Smithers. 2015. "From barriers to limits to climate change adaptation: Path dependency and the speed of change." *Ecology and Society* 20(3): 5. http://dx.doi.org/10.5751/ES-07698-200305
43. IPCC 2022.
44. Keenan, Jesse M. 2014. "Material and social construction: A framework for the adaptation of buildings." *Enquiry: The ARCC Journal for Architectural Research* 11(1): 15. https://doi.org/10.17831/enq:arcc.v11i1.271
45. Keenan 2014.
46. Keenan, Jesse M., Benjamin D. Trump, William Hynes, and Igor Linkov. 2021. "Exploring the convergence of resilience processes and sustainable outcomes in post-COVID, post-Glasgow economies." *Sustainability* 13(23): 13415. https://doi.org/10.3390/su132313415
47. Adger et al. 2005.
48. IPCC 2022.
49. Keenan 2018, "Types and forms of resilience in local planning in the U.S.: Who does what?"
50. Wu Jianguo and Tong Wu. 2012. "Ecological resilience as a foundation for urban design and sustainability." In *Resilience in Ecology and Urban Design: Linking Theory and Practice for Sustainable Cities*, edited by Steward T. A. Pickett, Mary L. Cadenasso, and Brian McGrath, 211–229. Dordrecht: Springer Netherlands.
51. Wilson, Geoff A. 2014. "Community resilience: Path dependency, lock-in effects and transitional ruptures." *Journal of Environmental Planning and Management* 57(1): 1–26. https://doi.org/10.1002/eet.1620
52. Schuller, Mark, and Julie K. Maldonado. 2016. "Disaster capitalism." *Annals of Anthropological Practice* 40(1): 61–72. https://doi.org/10.1111/napa.12088
53. Keenan, Jesse M. 2018. "From climate change to national security: Analysis of the Obama administration's federal resilience mandates and measures." *Natural Hazards Review* 19(1): 04017022. https://doi.org/10.1061/(ASCE)NH.1527-6996.0000273; Keenan, Jesse M., Benjamin Trump, Eero Kytömaa, Gitanjali Adlakha-Hutcheon, and Igor Linkov. 2024. "The role of science in resilience planning for military-civilian domains in the U.S. and NATO." *Defence Studies*. https://doi.org/10.1080/14702436.2024.2365218
54. Tweed, Fiona, and Gordon Walker. 2011. "Some lessons for resilience from the 2011 multi-disaster in Japan." *Local Environment* 16(9): 937–942. https://doi.org/10.1080/13549839.2011.617949
55. Tweed and Walker 2011.
56. Hamin, Elisabeth M., and Nicole Gurran. 2009. "Urban form and climate change: Balancing adaptation and mitigation in the US and Australia." *Habitat International* 33(3): 238–245. https://doi.org/10.1016/j.habitatint.2008.10.005
57. Alexander, David E. 2013. "Resilience and disaster risk reduction: An etymological journey." *Natural Hazards and Earth System Sciences* 13(11): 2707–2716. https://doi.org/10.5194/nhess-13-2707-2013
58. Alexander 2013.
59. Alexander 2013.
60. Schwarz, Silke. 2018. "Resilience in psychology: A critical analysis of the concept." *Theory & Psychology* 28(4): 528–541. https://doi.org/10.1177/0959354318783584
61. Keenan, Jesse M., and Keely Maxwell. 2021. "Rethinking the design of resilience and adaptation indicators supporting coastal communities." *Journal of Environmental Planning and Management* 65(12): 2297–2317. https://doi.org/10.1080/09640568.2021.1971635
62. Scheffer, Marten, Stephen R. Carpenter, Vasilis Dakos, and Egbert H. van Nes. 2015. "Generic indicators of ecological resilience: Inferring the chance of a critical transition." *Annual Review of Ecology, Evolution, and Systematics* 46(1): 145–167. https://doi.org/10.1146/annurev-ecolsys-112414-054242
63. Keenan, Jesse M. 2019. *Climate Adaptation Finance and Investment in California*. London: Taylor & Francis.
64. Matarrita-Cascante, David, Bernardo Trejos, Hua Qin, Dongoh Joo, and Sigrid Debner. 2017. "Conceptualizing community resilience: Revisiting conceptual distinctions." *Community Development* 48(1): 105–123. https://doi.org/10.1080/15575330.2016.1248458
65. Koliou, Maria, John W. van de Lindt, Therese P. McAllister, Bruce R. Ellingwood, Maria Dillard, and Harvey Cutler. 2020. "State of the research in community resilience: Progress and challenges." *Sustainable and Resilient Infrastructure* 5(3): 131–151. https://doi.org/10.1080/23789689.2017.1418547

66. Leitner, Helga, Eric Sheppard, Sophie Webber, and Emma Colven. 2018. "Globalizing urban resilience." *Urban Geography* 39(8): 1276–1284. https://doi.org/10.1080/02723638.2018.1446870
67. Keenan, Jesse M. 2016, February 19th. "The resilience of resilience thinking: A critical trajectory." Presented at the Rebuild by Design University Conference, Columbia University, New York, NY.
68. Wu and Wu 2012.
69. Van den Berg, Hanne J., and Jesse M. Keenan. 2019. "Dynamic vulnerability in the pursuit of just adaptation processes: A Boston case study." *Environmental Science & Policy* 94: 90–100. https://doi.org/10.1016/j.envsci.2018.12.015
70. The field of journalism has also lacked a commitment to adaptation; today's journalists are the first generation to cover this beat.
71. Khan, Mizan R., and J. Timmons Roberts. 2013. "Adaptation and international climate policy." *Wiley Interdisciplinary Reviews: Climate Change* 4(3): 171–189. https://doi.org/10.1002/wcc.212
72. Pindyck, Robert S. 2017. "The use and misuse of models for climate policy." *Review of Environmental Economics and Policy* 11:1. https://doi.org/10.1093/reep/rew012
73. Kinley, Richard. 2017. "Climate change after Paris: From turning point to transformation." *Climate Policy* 17(1): 9–15. https://doi.org/10.1080/14693062.2016.1191009
74. Keenan 2018, "From climate change to national security: Analysis of the Obama administration's federal resilience mandates and measures.".
75. Mechler, Reinhard, Chandni Singh, Kristie Ebi, Riyante Djalante, Adelle Thomas, Rachel James, and Petra Tschakert. 2020. "Loss and damage and limits to adaptation: Recent IPCC insights and implications for climate science and policy." *Sustainability Science* 15: 1245–1251. https://doi.org/10.1007/s11625-020-00807-9
76. National Aeronautics and Space Administration (NASA) and U.S. Department of the Interior. 2022. "Interagency Forum on Climate Risks, Impacts and Adaptation." Washington, D.C.: NASA. https://www.fedcenter.gov/programs/greenhouse/ccforum/
77. Hill, Alice C., and Leonardo Martinez-Diaz. 2020. *Building a Resilient Tomorrow: How to Prepare for the Coming Climate Disruption*. New York: Oxford University Press.
78. Keenan 2018, "From climate change to national security: Analysis of the Obama administration's federal resilience mandates and measures."
79. National Institute of Standards and Technology (NIST). 2015, November 13th. "Four agencies join with NIST to co-sponsor new community resilience panel." Gaithersburg, MD: National Institute for Standards and Technology. https://www.nist.gov/news-events/news/2015/11/four-agencies-join-nist-co-sponsor-new-community-resilience-panel
80. White House. 2013, February 12th. "Statements and releases: Presidential policy directive—critical infrastructure security and resilience." Washington, D.C.: National Archives. https://obamawhitehouse.archives.gov/the-press-office/2013/02/12/presidential-policy-directive-critical-infrastructure-security-and-resil
81. White House 2013.
82. White House. 2014, October 31st. "Statements and releases: Obama Administration releases federal agency climate plans on fifth anniversary of Presidential Sustainability Initiative." Washington, D.C.: National Archives. https://obamawhitehouse.archives.gov/administration/eop/ceq/Press_Releases/_October_31_2014
83. Revkin, Andrew. 2017, March 14th. "Trump's defense chief cites climate change as national security challenge." *Science*. https://doi.org/10.1126/science.aal0911
84. See note 53; Keenan 2018, "From climate change to national security: Analysis of the Obama administration's federal resilience mandates and measures."
85. White House. 2024. "Justice40: A whole-of-government initiative." Washington, D.C.: Executive Office of the President of the United States. https://bidenwhitehouse.archives.gov/environmentaljustice/justice40/
86. Costley, Drew. 2022, January 6th. "Environmental justice in spotlight as WH official departs." Associated Press. https://apnews.com/article/climate-joe-biden-science-environment-and-nature-environment-a5eeb966bca1aba76c4d2c5a0706b610
87. Frank, Thomas. 2023, January 24th. "How the White House found EJ areas without using race." *E&E News*. https://www.eenews.net/articles/how-the-white-house-found-ej-areas-without-using-race/
88. Conley, Shannon, David M. Konisky, and Megan Mullin. 2022. "Delivering on environmental justice? US state implementation of the Justice40 initiative." *Publius: The Journal of Federalism* 53(3): 349–377. https://doi.org/10.1093/publius/pjad018

89. Energy Information Administration (EIA). 2022. "Residential energy consumption survey (RECS): Analysis & projections." Washington, D.C.: U.S. Department of Energy. https://www.eia.gov/consumption/residential/reports/2015/energybills/
90. National Centers for Environmental Information. 2025. "Billion-dollar weather and climate disasters." Asheville, NC: National Ocean Graphic and Atmospheric Administration. https://www.ncei.noaa.gov/access/billions/
91. Newman, Rebecca, and Ilan Noy. 2023. "The global costs of extreme weather that are attributable to climate change." *Nature Communications* 14(1): 6103. https://doi.org/10.1038/s41467-023-41888-1
92. Government Accountability Office (GAO). 2023, May 17th. "FEMA: Opportunity to strengthen management and address increasing challenges." GAO-23-106840. Washington, D.C.: U.S. Congress.
93. GAO 2023.
94. Bittle, Jake. 2023, September 29th. "Disaster recovery projects stall nationwide as FEMA runs out of money." *Grist*. https://grist.org/extreme-weather/fema-disaster-relief-fund-government-shutdown-recovery-congress/
95. Department of Homeland Security (DHS). 2008. "National response framework." Washington, D.C.: Department of Homeland Security; Department of Homeland Security (DHS). 2016. "National Disaster recovery framework." Washington, D.C.: Department of Homeland Security.
96. Robert T. Stafford Disaster Relief and Emergency Assistance Act (1988).
97. Federal Emergency Management Agency (FEMA). 2022. "Mitigation assistance: Building resilience infrastructure and communities." FEMA Policy FP-104-008-05. Washington, D.C.: Department of Homeland Security.
98. Adler, Dena P., and Emma Gosliner. 2019, September. *State Hazard Mitigation Plans & Climate Change: Rating the States 2019 Updated*. New York: Sabin Center for Climate Law, Columbia Law School, Columbia University.
99. Federal Emergency Management Agency (FEMA). 2023, July. *FEMA Response and Recovery Climate Change Planning Guidance*. Washington, D.C.: Department of Homeland Security and National Oceanographic Atmospheric Administration. https://www.fema.gov/sites/default/files/documents/fema_response-recovery_climate-change-planning-guidance_20230630.pdf
100. Flavelle, Christopher. 2021, December 3rd. "Billions for climate protection fuel new debate: Who deserves it most." *The New York Times*. https://www.nytimes.com/2021/12/03/climate/climate-change-infrastructure-bill.html
101. Emrich, Christopher T., Sanam K. Aksha, and Yao Zhou. 2022. "Assessing distributive inequities in FEMA's disaster recovery assistance fund allocation." *International Journal of Disaster Risk Reduction* 74: 102855. https://doi.org/10.1016/j.ijdrr.2022.102855
102. Federal Emergency Management Agency (FEMA). 2024. "Community Disaster Resilience Zones." Washington, D.C.: Department of Homeland Security. https://www.fema.gov/partnerships/community-disaster-resilience-zones
103. Congressional Research Service. 2024, January 22nd. "The Disaster Relief Fund: Overview and issues." R454484. Washington, D.C.: Congressional Research Service. https://crsreports.congress.gov/product/pdf/R/R45484
104. Weber, Anna. 2023, February 8th. "The BRIC wall: Capacity gaps put FEMA grants out of reach." New York: National Resources Defense Council. https://www.nrdc.org/bio/anna-weber/bric-wall-capacity-gaps-put-fema-grants-out-reach
105. Congressional Research Service. 2022, April 8th. "Flood buyouts: Federal funding for property acquisition." IN11911. Version 2. Washington, D.C.: Congressional Research Service. https://crsreports.congress.gov/product/pdf/IN/IN11911/2
106. U.S. Army Corps of Engineers (USACE). 2015, December 15th. "Planning bulletin: Clarification of existing policy for USACE participation in nonstructural flood risk management and coastal storm damage reduction measures." No. PB-2016-01. Washington, D.C.: Department of Defense. https://planning.erdc.dren.mil/toolbox/library/pb/PB2016_01.pdf
107. Cheng, Fanilla. 2021, November 17th. "Is compulsory managed retreat our future? Examining the U.S. Army Corps of Engineers' eminent domain policy." Brief. Washington, D.C.: New America.

108. Mach, Katharine J., Caroline M. Kraan, Miyuki Hino, A. R. Siders, Erica M. Johnston, and Christopher B. Field. 2019. "Managed retreat through voluntary buyouts of flood-prone properties." *Science Advances* 5(10): eaax8995. https://doi.org/10.1126/sciadv.aax8995
109. Greer, Alex, Sherri Brokopp Binder, and Elyse Zavar. 2022. "From hazard mitigation to climate adaptation: A review of home buyout program literature." *Housing Policy Debate* 32(1): 152–170. https://doi.org/10.1080/10511482.2021.1931930
110. Siders, Anne R. 2019. "Social justice implications of US managed retreat buyout programs." *Climatic Change* 152(2): 239–257. https://doi.org/10.1007/s10584-018-2272-5
111. McGhee, Devon J., Sherri Brokopp Binder, and Elizabeth A. Albright. 2020. "First, do no harm: Evaluating the vulnerability reduction of post-disaster home buyout programs." *Natural Hazards Review* 21(1): 05019002. https://doi.org/10.1061/(ASCE)NH.1527-6996.0000337
112. Vilay, Erica, and Phil Pollman. 2020. *Floodway Buyout Strategy for a Resilient Houston: A Systems Approach to Breaking the Dangerous and Expensive Cycle of Rebuilding in the Floodway.* Cambridge, MA: Kennedy School of Government, Harvard University. https://www.jchs.harvard.edu/sites/default/files/harvard_jchs_floodway_buyout_strategy_houston_vilay_pollman_2020_red.pdf
113. McGhee et al. 2020.
114. Dundon, Leah A., and Janey S. Camp. 2021. "Climate justice and home-buyout programs: Renters as a forgotten population in managed retreat actions." *Journal of Environmental Studies and Sciences* 11(3): 420–433. https://doi.org/10.1007/s13412-021-00691-4
115. Keeler, Andrew G., Megan Mullin, Dylan E. McNamara, and Martin D. Smith. 2022. "Buyouts with rentbacks: A policy proposal for managing coastal retreat." *Journal of Environmental Studies and Sciences* 12(3): 646–651. https://doi.org/10.1007/s13412-022-00762-0
116. Freudenberg, Robert, Ellis Calvin, Laura Tolkoff, and Dare Brawley. 2016. "Buy-in for buyouts: The case for managed retreat from flood zones." Cambridge, MA: Lincoln Institute of Land Policy.
117. Federal Emergency Management Agency (FEMA). 2024. "FEMA efforts advancing community-driven relocation." Washington, D.C.: Department of Homeland Security. https://www.fema.gov/fact-sheet/fema-efforts-advancing-community-driven-relocation
118. National Academies of Sciences, Engineering, and Medicine. 2024. *Community-Driven Relocation: Recommendations for the U.S. Gulf Coast and Beyond.* Washington, D.C.: National Academies Press.
119. Grade, A. M., A. R. Crimmins, S. Basile, M. R. Essig, L. Goldsmith, T. K. Maycock, A. McCarrick, A. Lustig, and A. Scheetz. 2023. "Appendix 5: Glossary." In *Fifth National Climate Assessment.* Crimmins, A. R., C. W. Avery, D. R. Easterling, K. E. Kunkel, B. C. Stewart, and T. K. Maycock, Eds. Washington, D.C.: U.S. Global Change Research Program. https://doi.org/10.7930/NCA5.2023.A5
120. Department of Housing and Urban Development (HUD). 2023. *Climate Resilience Implementation Guide: Community Driven Relocation.* Washington, D.C.: Department of Housing and Urban Development.
121. Olander, Lydia [White House Council on Environment Quality]. Email correspondence. May 16, 2023.
122. Buy-In Community Planning. 2024. https://buy-in.org/
123. Government Accountability Office (GAO). 2020, August 5th. "Climate change: A climate migration pilot program could enhance the nation's resilience and reduce federal fiscal exposure." GAO 20-488. Washington, D.C.: U.S. Congress. https://www.gao.gov/products/gao-20-488
124. White House Council on Environmental Quality (CEQ). Meeting on community-driven relocation. July 9, 2023.
125. The White House. 2021, January 27th. Executive Order 14008, "Tackling the Climate Crisis at Home and Abroad." Washington, D.C.: Executive Office of the President.
126. Office of the Federal Chief Sustainability Officer, Council on Environmental Quality (CEQ). 2024. *Progress: Federal Progress, Plans, and Performance; Agency-Specific Progress on Sustainability and Climate Resilience Goals.* Washington, D.C.: Executive Office of the President.
127. Department of Agriculture (USDA). 2024. *Climate Adaptation Plan: 2024–2027.* Washington, D.C.: Department of Agriculture.

128. Department of the Interior (DOI). 2024. *Department of the Interior Climate Adaptation Plan: 2024*. Washington, D.C.: Department of the Interior.
129. Department of Commerce. 2024. *2024–2027 Climate Adaptation Plan*. Washington, D.C.: Department of Commerce.
130. Department of Defense. 2024. *2024–2027 Climate Adaptation Plan*. Arlington, VA: Department of Defense.
131. National Aeronautics and Space Administration (NASA). 2024. *NASA's Climate Adaptation Plan*. Washington, D.C.: National Aeronautics and Space Administration.
132. U.S. Army Corps of Engineers (USACE). 2024. *Climate Preparedness and Resilience: Policy Statement*. Arlington, VA: Department of Defense.
133. Department of Homeland Security (DHS). *Fiscal Years 2024–2027 Climate Adaptation Plan: Approach to Address Addressing Climate Hazard Impacts and Exposure*. Washington, D.C.: Department of Homeland Security.
134. Department of Homeland Security (DHS). "Updates to floodplain management and protection of wetlands regulations to implement the federal Flood Risk Management Standard." 89 FR 56929. Washington, D.C.: Federal Register. https://www.federalregister.gov/documents/2024/07/11/2024-15169/updates-to-floodplain-management-and-protection-of-wetlands-regulations-to-implement-the-federal
135. After the devastating cuts in the Trump administration, the EPA struggled to build the staff necessary to support more sustained APP development.
136. Environmental Protection Agency (EPA). 2024. *2024–2027 Climate Adaptation Plan*. Washington, D.C.: Environmental Protection Agency.
137. Department of Commerce. 2024. *2024–2027 Climate Adaptation Plan*. Washington, D.C.: Department of Commerce.
138. Department of Housing and Urban Development (HUD). 2024. *HUD 2024–2027 Federal Climate Adaptation Plan*. Washington, D.C.: Department of Housing and Urban Development.
139. General Services Administration (GSA). 2024. *2024–2027 Climate Change Risk Management Plan*. Washington, D.C.: General Services Administration.
140. Department of Energy (DOE). 2024. *U.S. Department of Energy Climate Adaptation Plan: 2024–2027*. Washington, D.C.: Department of Energy.
141. Department of Transportation (DOT). 2024. *2024–2027 Climate Adaptation Plan*. Washington, D.C.: Department of Transportation.
142. Social Security Administration (SSA). 2024. *2024–2027 Climate Adaptation Plan*. Washington, D.C.: Social Security Administration.
143. Social Security Administration (SSA). 2021. *2021 Climate Action Plan*. Washington, D.C.: Social Security Administration.
144. White House. 2023, September. "National Climate Resilience Framework." Washington, D.C.: Executive Office of the President.
145. United Nations Framework Convention on Climate Change (UNFCC). 2021. "Key aspects of the Paris Accord." Bonn: United Nations Framework Convention on Climate Change. https://unfccc.int/most-requested/key-aspects-of-the-paris-agreement
146. United States of America. 2025. *U.S. National Adaptation and Resilience Planning Strategy* Submitted as National Adaptation Plan (NAP). Bonn: United Nations Framework Convention on Climate Change. https://unfccc.int/documents/645358
147. Council of Economic Advisors. 2023, March. *Economic Report of the President*. Washington, D.C.: Executive Office of the President.
148. Urban Institute. 2022. "State and Local Expenditures." Washington, D.C.: Urban Institute. https://www.urban.org/policy-centers/cross-center-initiatives/state-and-local-finance-initiative/state-and-local-backgrounders/state-and-local-expenditures
149. Koski, Chris, and Megan Keating. 2018. "Holding back the storm: Target populations and state climate adaptation planning in America." *Review of Policy Research* 35(5): 691–716. https://doi.org/10.1111/ropr.12308
150. Dowds, Jonathan, and Lisa Aultman-Hall. 2015. "Barriers to implementation of climate adaptation frameworks by state departments of transportation." *Transportation Research Record* 2532(1): 21–28.https://doi.org/10.3141/2532-03
151. Molinaroli, Emanuela, Stefano Guerzoni, and Daniel Suman. 2019. "Do the adaptations of Venice and Miami to sea level rise offer lessons for other vulnerable coastal cities?" *Environmental Management* 64(4): 391–415. https://doi.org/10.1007/s00267-019-01198-z

152. Forerunner. 2024. https://www.withforerunner.com/
153. Keenan 2019.
154. Keenan 2019.
155. Goonesekera, Sascha M., and Marta Olazabal. 2022. "Climate adaptation indicators and metrics: State of local policy practice." *Ecological Indicators* 145: 109657.https://doi.org/10.1016/j.ecolind.2022.109657
156. Cvitanovic, Christopher, Mark Howden, R. M. Colvin, Albert Norström, Alison M. Meadow, and P. F. E. Addison. 2019. "Maximising the benefits of participatory climate adaptation research by understanding and managing the associated challenges and risks." *Environmental Science & Policy* 94: 20-31. https://doi.org/10.1016/j.envsci.2018.12.028
157. Uittenbroek, Caroline J., Heleen L. P. Mees, Dries L. T. Hegger, and Peter P. J. Driessen. 2019. "The design of public participation: Who participates, when and how? Insights in climate adaptation planning from the Netherlands." *Journal of Environmental Planning and Management* 62(14): 2529–2547. https://doi.org/10.1080/09640568.2019.1569503
158. Shearer, Christine. 2012. "The political ecology of climate adaptation assistance: Alaska Natives, displacement, and relocation." *Journal of Political Ecology* 19(1): 174–183. https://doi.org/10.2458/v19i1.21725
159. Nordgren, John, Missy Stults, and Sara Meerow. 2016. "Supporting local climate change adaptation: Where we are and where we need to go." *Environmental Science & Policy* 66: 344–352.
160. Lyles, Ward, Philip Berke, and Kelly Heiman Overstreet. 2018. "Where to begin municipal climate adaptation planning? Evaluating two local choices." *Journal of Environmental Planning and Management* 61(11): 1994–2014. https://doi.org/10.1080/09640568.2017.1379958
161. Adler and Gosliner 2019; Hu, Qiao, Zhenghong Tang, Lei Zhang, Yuanyuan Xu, Xiaolin Wu, and Ligang Zhang. 2018. "Evaluating climate change adaptation efforts on the US 50 states' hazard mitigation plans." *Natural Hazards* 92: 783–804. https://doi.org/10.1007/s11069-018-3225-z
162. Center for Climate and Energy Studies. 2023. "U.S. State Climate Action Plans." Arlington, VA: Center for Climate and Energy Studies. https://www.c2es.org/document/climate-action-plans/
163. Environmental Protection Agency (EPA). 2024, March 11th. "45 states, large metro areas submit climate action plans under President Biden's Inflation Reduction Act." Washington, D.C.: Environmental Protection Agency. https://www.epa.gov/newsreleases/45-states-large-metro-areas-submit-climate-action-plans-under-president-bidens
164. Rai, Saatvika. 2020. "Policy adoption and policy intensity: Emergence of climate adaptation planning in US states." *Review of Policy Research* 37(4): 444–463. https://doi.org/10.1111/ropr.12383
165. Georgetown Climate Center. 2023. State Adaptation Progress Tracker. Washington, D.C.: Georgetown Law, Georgetown University. https://www.georgetownclimate.org/adaptation/plans.html
166. Bierbaum, Rosina, Joel B. Smith, Arthur Lee, Maria Blair, Lynne Carter, F. Stuart Chapin, and Paul Fleming. 2013. "A comprehensive review of climate adaptation in the United States: More than before, but less than needed." *Mitigation and Adaptation Strategies for Global Change* 18: 361–406. https://doi.org/10.1007/s11027-012-9423-1
167. Miao, Qing. 2019. "What affects government planning for climate change adaptation: Evidence from the US states." *Environmental Policy and Governance* 29(5): 376–394. https://doi.org/10.1002/eet.1866
168. Smaldone, Amy, and Mark L. J. Wright. 2024, May 14th. "Local governments in the U.S.: A breakdown by number and type." St. Louis, MO: Federal Reserve Bank of St. Louis. https://www.stlouisfed.org/publications/regional-economist/2024/march/local-governments-us-number-type
169. Shi, Linda, Eric Chu, and Jessica Debats. 2015. "Explaining progress in climate adaptation planning across 156 US municipalities." *Journal of the American Planning Association* 81(3): 191–202. https://doi.org/10.1080/01944363.2015.1074526
170. Fiack, Duran, Jeremy Cumberbatch, Michael Sutherland, and Nadine Zerphey. 2021. "Sustainable adaptation: Social equity and local climate adaptation planning in US cities." *Cities* 115: 103235.https://doi.org/10.1016/j.cities.2021.103235

171. Shi, Linda. 2019. "Promise and paradox of metropolitan regional climate adaptation." *Environmental Science & Policy* 92: 262–274. https://doi.org/10.1016/j.envsci.2018.11.002
172. Keenan 2018, "Types and forms of resilience in local planning in the U.S.: Who does what?"
173. Tietjen, Bethany, Jenna Clark, and Erin Coughlan de Perez. 2024. "Progress and gaps in US adaptation policy at the local level." *Global Environmental Change* 87: 102882. https://doi.org/10.1016/j.gloenvcha.2024.102882
174. Reckien, Diana, Attila Buzasi, Marta Olazabal, Niki-Artemis Spyridaki, Peter Eckersley, Sofia G. Simoes, and Monica Salvia. 2023. "Quality of urban climate adaptation plans over time." *Urban Sustainability* 3(1): 13. https://doi.org/10.1038/s42949-023-00085-1
175. Hamin, Elisabeth M., Nicole Gurran, and Ana Mesquita Emlinger. 2014. "Barriers to municipal climate adaptation: Examples from coastal Massachusetts' smaller cities and towns." *Journal of the American Planning Association* 80(2): 110–122. https://doi.org/10.1080/01944363.2014.949590; Becker, Austin, and Eric Kretsch. 2019. "The leadership void for climate adaptation planning: Case study of the port of providence (Rhode Island, United States)." *Frontiers in Earth Science* 7: 29. https://doi.org/10.3389/feart.2019.00029

Chapter 4

1. Warren, Elizabeth. 2019, January 21st. Letter to the CEOs of Bank of America, Bank of New York Mellon, Citigroup, Goldman Sachs, J.P. Morgan Chase & Co., Morgan Stanley & Co., State Street Corporation, and Wells Fargo.
2. Bin, Okmyung, and Craig E. Landry. 2013. "Changes in implicit flood risk premiums: Empirical evidence from the housing market." *Journal of Environmental Economics and Management* 65(3): 361–376. https://doi.org/10.1016/j.jeem.2012.12.002
3. Wang, Yuhan, and David J. Lewis. 2024. "Wildfires and climate change have lowered the economic value of western US forests by altering risk expectations." *Journal of Environmental Economics and Management* 123(1): 102894. https://doi.org/10.1016/j.jeem.2023.102894
4. Fonner, Robert, Germán Izón, Blake E. Feist, and Katie Barnas. 2023. "Capitalization of reduced flood risk into housing values following a floodplain restoration investment." *Journal of Environmental Economics and Policy* 12(3): 305–323. https://doi.org/10.1080/21606544.2022.2136765
5. Fonner et al. 2023.
6. Bin, Okmyung, Thomas W. Crawford, Jamie B. Kruse, and Craig E. Landry. 2008. "Viewscapes and flood hazard: Coastal housing market response to amenities and risk." *Land Economics* 84(3): 434–448. https://doi.org/10.3368/le.84.3.434
7. Jin, Di, Porter Hoagland, Donna K. Au, and Jun Qiu. 2015. "Shoreline change, seawalls, and coastal property values." *Ocean & Coastal Management* 114: 185–193. https://doi.org/10.1016/j.ocecoaman.2015.06.025
8. Stetler, Kyle M., Tyron J. Venn, and David E. Calkin. 2010. "The effects of wildfire and environmental amenities on property values in northwest Montana, USA." *Ecological Economics* 69(11): 2233–2243. https://doi.org/10.1016/j.ecolecon.2010.06.009
9. Calil, Juliano, Geraldine Fauville, Anna Carolina Muller Queiroz, Kelly L. Leo, Alyssa G. Newton Mann, Tiffany Wise-West, Paulo Salvatore, and Jeremy N. Bailenson. 2021. "Using virtual reality in sea level rise planning and community engagement—an overview." *Water* 13(9): 1142. https://doi.org/10.3390/w13091142
10. Treuer, Galen, Kenneth Broad, and Robert Meyer. 2018. "Using simulations to forecast homeowner response to sea level rise in South Florida: Will they stay or will they go?" *Global Environmental Change* 48: 108–118. https://doi.org/10.1016/j.gloenvcha.2017.10.008
11. Bernstein, Asaf, Matthew T. Gustafson, and Ryan Lewis. 2019. "Disaster on the horizon: The price effect of sea level rise." *Journal of Financial Economics* 134(2): 253–272. https://doi.org/10.1016/j.jfineco.2019.03.013
12. Bernstein 2019.
13. Bernstein 2019.
14. Bernstein 2019.
15. Bernstein 2019.
16. Keys, Benjamin J., and Philip Mulder. 2020. "Neglected no more: Housing markets, mortgage lending, and sea level rise." Working Paper No. 27930. Cambridge, MA: National Bureau of Economic Research. https://www.nber.org/papers/w27930

17. Barrage, Lint, and Jacob Furst. 2019. "Housing investment, sea level rise, and climate change beliefs." *Economics Letters* 177:105–108. https://doi.org/10.1016/j.econlet.2019.01.023
18. Parton, Lee C., and Steven J. Dundas. 2020. "Fall in the sea, eventually? A green paradox in climate adaptation for coastal housing markets." *Journal of Environmental Economics and Management* 104: 102381. https://doi.org/10.1016/j.jeem.2020.102381
19. Duany, Andres, Elizabeth Plater-Zyberk, and Jeff Speck. 2020. *Suburban Nation: The Rise of Sprawl and the Decline of the American Dream.* New York: North Point Press.
20. Baldauf, Markus, Lorenzo Garlappi, and Constantine Yannelis. 2020. "Does climate change affect real estate prices? Only if you believe in it." *The Review of Financial Studies* 33(3): 1256–1295. https://doi.org/10.1093/rfs/hhz073
21. Baldauf et al. 2020.
22. Bernstein et al. 2019.
23. Folkers, Andreas. 2024. "Calculative futures between climate and finance: A tragedy of multiple horizons." *The Sociological Review*. https://doi.org/10.1177/00380261241258832
24. Javeline, Debra, Tracy Kijewski-Correa, and Angela Chesler. 2019. "Does it matter if you 'believe' in climate change? Not for coastal home vulnerability." *Climatic Change* 155: 511–532. https://doi.org/10.1007/s10584-019-02513-7
25. Goldberg, Matthew H., Sander van der Linden, Anthony Leiserowitz, and Edward Maibach. 2020. "Perceived social consensus can reduce ideological biases on climate change." *Environment and Behavior* 52(5): 495–517. https://doi.org/10.1177/0013916519853302
26. Goldberg, Matthew H., Abel Gustafson, Seth A. Rosenthal, and Anthony Leiserowitz. 2021. "Shifting Republican views on climate change through targeted advertising." *Nature Climate Change* 11(7): 573–577. https://doi.org/10.1038/s41558-021-01070-1
27. Treuer et al. 2018.
28. Beck, Jason, and Meimei Lin. 2020. "The impact of sea level rise on real estate prices in coastal Georgia." *Review of Regional Studies* 50(1): 43–52.
29. Ma, Lala, Margaret Walls, Matthew Wibbenmeyer, and Connor Lennon. 2024. "Risk disclosure and home prices: Evidence from California wildfire hazard zones." *Land Economics* 100(1): 6–21. https://doi.org/10.3368/le.100.1.102122-0087R
30. Hino, Miyuki, and Marshall Burke. 2021. "The effect of information about climate risk on property values." *Proceedings of the National Academy of Sciences* 118(17): e2003374118. https://doi.org/10.1073/pnas.2003374118
31. Bakkensen, Laura A., and Lala Ma. 2020. "Sorting over flood risk and implications for policy reform." *Journal of Environmental Economics and Management* 104: 102362. https://doi.org/10.1016/j.jeem.2020.102362
32. Bakkensen and Ma 2020.
33. Ray, Anne, Ruoniu Wang, Diep Nguyen, Jim Martinez, Nicholas Taylor, and Jennison Kipp Searcy. 2019. "Household energy costs and the housing choice voucher program: Do utility allowances pay the bills?" *Housing Policy Debate* 29(4): 607–626. https://doi.org/10.1080/10511482.2019.1566158
34. Gakpetor, Godfred, and Maggie Shober. 2023. "Analysis of select utilities passing fuel costs and risks onto customers." *Environmental Progress & Sustainable Energy* 42(4): e14167. https://doi.org/10.1002/ep.14167
35. American Council for an Energy-Efficient Economy. 2019. "How energy efficiency can help low-income households in Florida." Washington, D.C.: American Council for an Energy-Efficient Economy. https://www.aceee.org/sites/default/files/pdf/fact-sheet/ses-florida-100917.pdf
36. Brewer, Dylan, and Sarah Goldgar. 2024. "The effect of extreme temperatures on evictions." *Journal of Environmental Economics and Management* 128: 103055. https://doi.org/10.1016/j.jeem.2024.103055
37. See note 11.
38. Fu, Xinyu, and Jan Nijman. 2021. "Sea level rise, homeownership, and residential real estate markets in South Florida." *The Professional Geographer* 73(1): 62–71. https://doi.org/10.1080/00330124.2020.1818586
39. Fu and Nijman 2021.
40. Braswell, Anna E., Stefan Leyk, Dylan S. Connor, and Johannes H. Uhl. 2022. "Creeping disaster along the US coastline: Understanding exposure to sea level rise and hurricanes through historical development." *PLOS One* 17(8): e0269741. https://doi.org/10.1371/journal.pone.0269741

41. Siders, A. R., and Jesse M. Keenan. 2020. "Variables shaping coastal adaptation decisions to armor, nourish, and retreat in North Carolina." *Ocean & Coastal Management* 183: 105023. https://doi.org/10.1016/j.ocecoaman.2019.105023
42. Walsh, Patrick, Charles Griffiths, Dennis Guignet, and Heather Klemick. 2019. "Adaptation, sea level rise, and property prices in the Chesapeake Bay watershed." *Land Economics* 95(1): 19–34. https://doi.org/10.3368/le.95.1.19
43. Bunten, Devin Michaelle, and Matthew E. Kahn. 2018. *The Impact of Emerging Climate Risks on Urban Real Estate Price Dynamics.* Working Paper No. 20018. Cambridge, MA: National Bureau of Economic Research. https://www.nber.org/papers/w20018
44. Katz, Lily. 2021, July 14th. "1 in 5 Americans believes climate change is hurting home values in their area: Survey." *Redfin News.* https://www.redfin.com/news/climate-change-impact-home-values-survey/
45. Katz 2021.
46. Giglio, Stefano, Matteo Maggiori, Krishna Rao, Johannes Stroebel, and Andreas Weber. 2021. "Climate change and long-run discount rates: Evidence from real estate." *The Review of Financial Studies* 34(8): 3527–3571. https://doi.org/10.1093/rfs/hhab032
47. Keenan, Jesse M., Thomas Hill, and Anurag Gumber. 2018. "Climate gentrification: From theory to empiricism in Miami-Dade County, Florida." *Environmental Research Letters* 13(5): 054001. https://doi.org/10.1088/1748-9326/aabb32
48. Tarui, Nori, Seth Urbanski, Quang Loc Lam, Makena Coffman, and Conrad Newfield. 2023. "Sea level rise risk interactions with coastal property values: A case study of Oʻahu, Hawaiʻi." *Climatic Change* 176(9):130. https://doi.org/10.1007/s10584-023-03602-4
49. Catma, Serkan. 2021. "The price of coastal erosion and flood risk: A hedonic pricing approach." *Oceans* 2(1): 149–161. https://doi.org/10.3390/oceans2010009
50. Murfin, Justin, and Matthew Spiegel. 2020. "Is the risk of sea level rise capitalized in residential real estate?" *The Review of Financial Studies* 33(3): 1217–1255. https://doi.org/10.1093/rfs/hhz134
51. Fu and Nijman 2021.
52. McNamara, Dylan E., Martin D. Smith, Zachary Williams, Sathya Gopalakrishnan, and Craig E. Landry. 2024. "Policy and market forces delay real estate price declines on the US coast." *Nature Communications* 15(1): 2209. https://doi.org/10.1038/s41467-024-46548-6
53. Tyndall, Justin. 2023. "Sea level rise and home prices: Evidence from Long Island." *The Journal of Real Estate Finance and Economics* 67(4): 579–605. https://doi.org/10.1007/s11146-021-09868-8
54. Hino and Burke 2021.
55. Contat, Justin, Carrie Hopkins, Luis Mejia, and Matthew Suandi. 2024. "When climate meets real estate: A survey of the literature." *Real Estate Economics* 52(3): 618–659. https://doi.org/10.1111/1540-6229.12489
56. Bakkensen, Laura A., and Lint Barrage. 2022. "Going underwater? Flood risk belief heterogeneity and coastal home price dynamics." *The Review of Financial Studies* 35(8): 3666–3709. https://doi.org/10.1093/rfs/hhab122
57. Gourevitch, Jesse D., Carolyn Kousky, Yanjun Liao, Christoph Nolte, Adam B. Pollack, Jeremy R. Porter, and Joakim A. Weill. 2023. "Unpriced climate risk and the potential consequences of overvaluation in US housing markets." *Nature Climate Change* 13(3): 250–257. https://doi.org/10.1038/s41558-023-01594-8
58. Katz, Lily, and Chen Zhao. 2024, February 28th. "The U.S. housing market gained $2 trillion in value over the last year." *Redfin News.* https://www.redfin.com/news/housing-market-value-december-2023/
59. Gourevitch et al. 2023.
60. Clayton, Jim, Steven Devaney, Sarah Sayce, and Jorn Van de Wetering. 2021. "Climate risk and real estate prices: What do we know?" *Journal of Portfolio Management* 47(10): 75–90. https://doi.org/10.3905/jpm.2021.1.278
61. Fisher, Jeffrey D., and Sara R. Rutledge. 2021. "The impact of hurricanes on the value of commercial real estate." *Business Economics* 56(3): 129–145. https://doi.org/10.1057/s11369-021-00212-9
62. Addoum, Jawad M., Piet Eichholtz, Eva Steiner, and Erkan Yönder. 2024. "Climate change and commercial real estate: Evidence from Hurricane Sandy." *Real Estate Economics* 52(3): 687–713. https://doi.org/10.1111/1540-6229.12435

63. Chiang, Chia-Chun, Zifeng Feng and David Harrison. 2022. "Heat stress and commercial real estate returns." SSRN Working Paper. http://dx.doi.org/10.2139/ssrn.4150066
64. Salisu, Afees A., Yinka S. Hammed, and Ibrahim Ngananga Ouattara. 2023. "Climate change, technology shocks and the US equity real estate investment trusts (REITs)." *Sustainability* 15(19): 14536. https://doi.org/10.3390/su151914536
65. Feng, Zifeng, Ran Lu-Andrews, and Zhonghua Wu. 2024. "Commercial real estate in the face of climate risk: Insights from REITs." *Journal of Real Estate Research.* https://doi.org/10.1080/08965803.2024.2352917
66. Sayce, Sarah Louise, Jim Clayton, Steven Devaney, and Jorn van de Wetering. 2022. "Climate risks and their implications for commercial property valuations." *Journal of Property Investment & Finance* 40(4): 430–443. https://doi.org/10.1108/JPIF-02-2022-0018
67. Sirmans, Stace, G. Stacy Sirmans, Greg Smersh, and Daniel T. Winkler. 2024. "Perceptions of climate change and the pricing of disaster risk in commercial real estate." SSRN Working Paper. https://dx.doi.org/10.2139/ssrn.4851196
68. Yildirim, Yildiray, and Bing Zhu. 2024. "Exploring climate risk, risk retention, and CMBS: Understanding their interplay." Working Paper. https://www.haas.berkeley.edu/wp-content/uploads/Yildirim-Yildiray-Paper-Pre-WFA-2024.pdf
69. Holtermans, Rogier, Matthew E. Kahn, and Nils Kok. 2024. "Climate risk and commercial mortgage delinquency." *Journal of Regional Science* 64(4): 994-1037. https://doi.org/10.1111/jors.12681
70. Holtermans et al. 2024.
71. Behal, Bob, and Brian Monteleone. 2023. "Commercial Real Estate: Weathering the Storm." New York: Vanguard Advisors. Click here to enter text.
72. Federal Reserve Bank of St. Louis. 2024. "FRED economic data: Financial accounts of the United States; Released tables; Mortgage debt outstanding, million of dollars; End of period." St. Louis, MO: Federal Reserve Bank of St. Louis. https://fred.stlouisfed.org/release/tables?eid=1192326&rid=52
73. Mortgage lenders may also face transition risks to the extent that local economic conditions may deteriorate in locations that are heavily dependent on fossil fuel–centered economic output and labor participation and are otherwise undiversified or unprepared to transition to a low-carbon economy.
74. Kousky, Carolyn, Mark Palim, and Ying Pan. 2020. "Flood damage and mortgage credit risk: A case study of Hurricane Harvey." *Journal of Housing Research* 29(1): S86–S120. https://doi.org/10.1080/10527001.2020.1840131
75. Kousky et al. 2020.
76. Nguyen, Duc Duy, Steven Ongena, Shusen Qi, and Vathunyoo Sila. 2022. "Climate change risk and the cost of mortgage credit." *Review of Finance* 26(6): 1509–1549. https://doi.org/10.1093/rof/rfac013
77. Nguyen et al. 2022.
78. Duanmu, Jun, Yongjia Li, Meimei Lin, and Salman Tahsin. 2022. "Natural disaster risk and residential mortgage lending standards." *Journal of Real Estate Research* 44(1): 106–130. https://doi.org/10.1080/08965803.2021.2013613
79. Duanmu et al. 2022.
80. Duan, Tinghua, and Frank Weikai Li. 2024. "Climate change concerns and mortgage lending." *Journal of Empirical Finance* 75: 101445. https://doi.org/10.1016/j.jempfin.2023.101445
81. Baranyai, Eszter, and Ádám Banai. 2022. "Heat projections and mortgage characteristics: Evidence from the USA." *Climatic Change* 175(3): 14. https://doi.org/10.1007/s10584-022-03465-1
82. Furfine, Craig. 2020. "The impact of risk retention regulation on the underwriting of securitized mortgages." *Journal of Financial Services Research* 58(2): 91–114. https://doi.org/10.1007/s10693-019-00308-6
83. Gete, Pedro, Athena Tsouderou, and Susan M. Wachter. 2024. "Climate risk in mortgage markets: Evidence from Hurricanes Harvey and Irma." *Real Estate Economics* 52(3): 660–686. https://doi.org/10.1111/1540-6229.12477
84. Gete et al. 2024.
85. Email correspondence. Amine Ouazad. April 4, 2024; Kahn, Matthew E., Amine Ouazad, and Erkan Yönder. 2024. "Adaptation using financial markets: Climate risk diversification through

securitization." Working Paper No. 32244. Cambridge, MA: National Bureau of Economic Research. https://www.nber.org/system/files/working_papers/w32244/w32244.pdf
86. Ibid.
87. Bakkensen, Laura, Toan Phan, and Russell Wong. 2023. "Leveraging the disagreement on climate change: Theory and evidence." Working Paper 23-01R. Richmond, VA: Federal Reserve Bank of Richmond. https://www.richmondfed.org/-/media/RichmondFedOrg/publications/research/working_papers/2023/wp23-01.pdf
88. Sadasivam, Naveena, and Clayton Aldern. 2021, August 11th. "Has Fannie Mae's $95 billion in green bonds made anything greener?" *Grist.* https://grist.org/accountability/fannie-mae-green-bond-building-program/
89. Keenan, Jesse M., and Jacob T. Bradt. 2020. "Underwaterwriting: From theory to empiricism in regional mortgage markets in the US." *Climatic Change* 162(4): 2043–2067. https://doi.org/10.1007/s10584-020-02734-1
90. Martinez-Diaz, Leonardo, and Jesse M. Keenan, eds. 2020. *Managing Climate Risk in the U.S. Financial System.* Washington, D.C.: U.S. Commodity Futures Trading Commission.
91. Baranyai, Eszter, and Ádám Banai. 2022. "Feeling the Heat: Mortgage Lending and Central Bank Options." *Financial and Economic Review* 21(1): 5–31. http://doi.org/10.33893/FER.21.1.5
92. Keenan and Bradt 2020.
93. Keenan and Bradt 2020.
94. Avtar, Ruchi, Kristian Blickle, Rajashri Chakrabarti, Janavi Janakiraman, and Maxim Pinkovskiy. (2021). *Understanding the Linkages between Climate Change and Inequality in the United States.* Staff Reports No. 991. New York: Federal Reserve Bank of New York. http://dx.doi.org/10.2139/ssrn.3961093
95. Montgomery, Brooklyn, and Monica Palmeira. 2023. "Bluelining: Climate financial discrimination on the horizon." Oakland, CA: Greenlining Institute. https://greenlining.org/publications/bluelining-climate-financial-discrimination-on-the-horizon/
96. Jacobson, Lindsay. 2021, September 20th. "Banks consider climate risk for home loans, a process called 'underwaterwriting' or 'blue-lining.'" CNBC. https://www.cnbc.com/2021/09/20/blue-lining-and-underwaterwriting-banks-consider-climate-change-risk.html
97. Linscott, Gabrielle, Andrea Rishworth, Brian King, and Mikael P. Hiestand. 2022. "Uneven experiences of urban flooding: Examining the 2010 Nashville flood." *Natural Hazards* 110: 629–653. https://doi.org/10.1007/s11069-021-04961-w
98. Liévanos, Raoul S. 2020. "Racialised uneven development and multiple exposure: Sea-level rise and high-risk neighbourhoods in Stockton, CA." *Cambridge Journal of Regions, Economy and Society* 13(2): 381–404. https://doi.org/10.1093/cjres/rsaa009
99. Bakkensen and Ma 2020.
100. The Equal Credit Opportunity Act (ECOA). 15 U.S.C. 1691, et seq.
101. Consumer Financial Protection Bureau (CFPB). 2013. "CFPB consumers laws and regulations: Equal Credit Opportunity Act (ECOA)." Washington, D.C.: Consumer Financial Protection Bureau. https://files.consumerfinance.gov/f/201306_cfpb_laws-and-regulations_ecoa-combined-june-2013.pdf
102. CFPB 2013.
103. Keenan, Jesse M., Elizabeth Mattiuizzi, and Donta Council. 2024. "Bridging community investment and resilience in the Community Reinvestment Act." In *What's Possible: Investing NOW for Prosperous, Sustainable Neighborhoods.* New York: Federal Reserve Bank of New York.
104. Keenan, Jesse M. 2023, July 26th. "Fiscal and economic implications of climate change on infrastructure systems in the United States." Testimony in the 118th Congress, U.S. Senate Budget Committee Hearing, "Beyond the breaking point: The fiscal consequences of climate change on infrastructure." Washington, D.C.: Library of Congress. This testimony is based, in part, on sections of this chapter that were in production at the time of the testimony.
105. Lin, Ning, Robert E. Kopp, Benjamin P. Horton, and Jeffrey P. Donnelly. 2016. "Hurricane Sandy's flood frequency increasing from year 1800 to 2100." *Proceedings of the National Academy of Sciences* 113(43): 12071–12075. https://doi.org/10.1073/pnas.1604386113
106. Chester, Mikhail V., B. Shane Underwood, and Constantine Samaras. 2020. "Keeping infrastructure reliable under climate uncertainty." *Nature Climate Change* 10(6): 488–490. https://doi.org/10.1038/s41558-020-0741-0

107. Contat et al. 2024.
108. Beirne, John, Nuobu Renzhi, and Ulrich Volz. 2021. "Bracing for the typhoon: Climate change and sovereign risk in Southeast Asia." *Sustainable Development* 29(3): 537–551. https://doi.org/10.1002/sd.2199; Beirne, John, Nuobu Renzhi, and Ulrich Volz. 2021. "Feeling the heat: Climate risks and the cost of sovereign borrowing." *International Review of Economics & Finance* 76: 920–936. https://doi.org/10.1016/j.iref.2021.06.019
109. Schlenker, Wolfram, and Charles A. Taylor. 2021. "Market expectations of a warming climate." *Journal of Financial Economics* 142(2): 627–640. https://doi.org/10.1016/j.jfineco.2020.08.019
110. Huynh, Thanh D., and Ying Xia. 2023. "Panic selling when disaster strikes: Evidence in the bond and stock markets." *Management Science* 69(12): 7448–7467. https://doi.org/10.1287/mnsc.2021.4018; Huynh, Thanh D., and Ying Xia. 2021. "Climate change news risk and corporate bond returns." *Journal of Financial and Quantitative Analysis* 56(6): 1985–2009. https://doi.org/10.1017/S0022109020000757
111. Venturini, Alessio. 2022. "Climate change, risk factors and stock returns: A review of the literature." *International Review of Financial Analysis* 79: 101934. https://doi.org/10.1016/j.irfa.2021.101934
112. Painter, Marcus. 2020. "An inconvenient cost: The effects of climate change on municipal bonds." *Journal of Financial Economics* 135(2): 468–482. https://doi.org/10.1016/j.jfineco.2019.06.006
113. Acharya, Viral V., Timothy Johnson, Suresh Sundaresan, and Tuomas Tomunen. 2022. "Is physical climate risk priced? Evidence from regional variation in exposure to heat stress." Working Paper No. 30445. Cambridge, MA: National Bureau of Economic Research. https://www.nber.org/papers/w30445?
114. Li, Dan, Romesh Saigal, and Peter Adriaens. 2024. "Impact of flooding and drought risks on the cost of bond financing for water utilities." *Environmental Science & Technology* 58(32): 14068–14077. https://doi.org/10.1021/acs.est.4c01863; Jeon, Woongchan, Lint Barrage, and Kieran James Walsh. 2025. "Pricing climate risks: Evidence from wildfires and municipal bonds." Economic Working Paper Series 25. Zurich: ETH Zurich, Center of Economic Research. https://doi.org/10.3929/ethz-b-000729307
115. Keenan, Jesse M. 2019. *Climate Adaptation Finance and Investment in California*. London: Taylor & Francis.
116. Goldsmith-Pinkham, Paul, Matthew T. Gustafson, Ryan C. Lewis, and Michael Schwert. 2023. "Sea-level rise exposure and municipal bond yields." *The Review of Financial Studies* 36(11): 4588–4635. https://doi.org/10.1093/rfs/hhad041
117. Acharya, Viral V., Timothy Johnson, Suresh Sundaresan, and Tuomas Tomunen. 2022. "Is physical climate risk priced? Evidence from regional variation in exposure to heat stress." Working Paper No. 30445. Cambridge, MA: National Bureau of Economic Research. https://www.nber.org/papers/w30445?
118. Goldsmith-Pinkham et al. 2023.
119. Shi, Linda, and Andrew M. Varuzzo. 2020. "Surging seas, rising fiscal stress: Exploring municipal fiscal vulnerability to climate change." *Cities* 100: 102658. https://doi.org/10.1016/j.cities.2020.102658
120. Shi and Varuzzo. 2020.
121. Shi, Linda, William Butler, Tisha Holmes, Ryan Thomas, Anthony Milordis, Jonathan Ignatowski, Yousuf Mahid, and Austin M. Aldag. 2024. "Can Florida's coast survive its reliance on development? Fiscal vulnerability and funding woes under sea level rise." *Journal of the American Planning Association* 90(2): 367–383. https://doi.org/10.1080/01944363.2023.2249866
122. Jerch, Rhiannon, Matthew E. Kahn, and Gary C. Lin. 2023. "Local public finance dynamics and hurricane shocks." *Journal of Urban Economics* 134: 103516. https://doi.org/10.1016/j.jue.2022.103516
123. Moody's Investor Service. 2022, April 21st. "U.S. municipal bond defaults and recoveries, 1970–2021."
124. Disruptions in project-based revenue may also delay revenue-sharing trigger events that provide a cash-flow upside to public entity partners in public-private infrastructure development.
125. Yu, Jinhai, and Zhiwei Zhang. 2023. "Local debt financing in the shadow of storms: Disrupted and destructed?" *Public Performance & Management Review* 46(4): 942–969. https://doi.org/10.1080/15309576.2023.2192943

126. Mousseau, John R. 2023, February 18th. "Florida takes municipalities out to the woodshed." Sarasota, FL: Cumberland Advisors. https://www.cumber.com/market-commentary/florida-takes-municipalities-out-woodshed
127. Keenan, Jesse M., Eric Chu, and Jacqueline Peterson. 2019. "From funding to financing: Perspectives shaping a research agenda for investment in urban climate adaptation." *International Journal of Urban Sustainable Development* 11(3): 297–308. https://doi.org/10.1080/19463138.2019.1565413
128. See generally, New York City Mayor's Office of Climate & Environmental Justice. 2022. "Climate resiliency design guidelines." Version 4.1. New York: City of New York. https://www.nyc.gov/assets/sustainability/downloads/pdf/publications/CRDG-4-1-May-2022.pdf; General Services Administration (GSA). 2022. "Facilities standards for the Public Buildings Service (P-100)." Washington, D.C.: U.S. General Services Administration.
129. Wang, Tianni, Zhuohua Qu, Zaili Yang, Timothy Nichol, Geoff Clarke, and Ying-En Ge. 2020. "Climate change research on transportation systems: Climate risks, adaptation and planning." *Transportation Research Part D: Transport and Environment* 88: 102553. https://doi.org/10.1016/j.trd.2020.102553
130. Gallo, Elizabeth M., Katie Spahr, Emily Grubert, and Terri S. Hogue. 2022. "Improving the decision-making process for stormwater management using life-cycle costs and a benefit analysis." *Journal of Sustainable Water in the Built Environment* 8(2): 04022001. https://doi.org/10.1061/JSWBAY.0000977
131. Kontokosta, Constantine E., Vincent J. Reina, and Bartosz Bonczak. 2020. "Energy cost burdens for low-income and minority households: Evidence from energy benchmarking and audit data in five US cities." *Journal of the American Planning Association* 86(1): 89–105. https://doi.org/10.1080/01944363.2019.1647446; Cardoso, Diego S., and Casey J. Wichman. 2022. "Water affordability in the United States." *Water Resources Research* 58(12): e2022WR032206. https://doi.org/10.1029/2022WR032206
132. Asghari, Vahid, Shu-Chien Hsu, and Hsi-Hsien Wei. 2021. "Expediting life cycle cost analysis of infrastructure assets under multiple uncertainties by deep neural networks." *Journal of Management in Engineering* 37(6): 04021059. https://doi.org/10.1061/(ASCE)ME.1943-5479.0000950; Qiao, Yaning, Joao Santos, Anne Stoner, and Gerardo Flinstch. 2020. "Climate change impacts on asphalt road pavement construction and maintenance: An economic life cycle assessment of adaptation measures in the State of Virginia, United States." *Journal of Industrial Ecology* 24(2): 342–355. https://doi.org/10.1111/jiec.12936
133. Lopez-Cantu, Tania, and Constantine Samaras. 2018. "Temporal and spatial evaluation of stormwater engineering standards reveals risks and priorities across the United States." *Environmental Research Letters* 13(7): 074006. https://doi.org/10.1088/1748-9326/aac696; Underwood, B. Shane, Giuseppe Mascaro, Mikhail V. Chester, Andrew Fraser, Tania Lopez-Cantu, and Constantine Samaras. 2020. "Past and present design practices and uncertainty in climate projections are challenges for designing infrastructure to future conditions." *Journal of Infrastructure Systems* 26(3): 04020026. https://doi.org/10.1061/(ASCE)IS.1943-555X.0000567
134. Neumann, James E., Paul Chinowsky, Jacob Helman, Margaret Black, Charles Fant, Kenneth Strzepek, and Jeremy Martinich. 2021. "Climate effects on US infrastructure: The economics of adaptation for rail, roads, and coastal development." *Climatic Change* 167(3): 44. https://doi.org/10.1007/s10584-021-03179-w. Note: A reduction in the useful life of an asset also represents a credit and/or capital risk to the extent that the asset may produce less revenue than was otherwise projected.
135. Miami-Dade County, Florida. 2021. "Miami-Dade County sea level rise strategy." Miami: Miami-Dade County, Florida. https://miami-dade-county-sea-level-rise-strategy-draft-mdc.hub.arcgis.com/
136. Jaglom, Wendy S., James R. McFarland, Michelle F. Colley, Charlotte B. Mack, Boddu Venkatesh, Rawlings L. Miller, and Juanita Haydel. 2014. "Assessment of projected temperature impacts from climate change on the US electric power sector using the Integrated Planning Model®." *Energy Policy* 73: 524–539. https://doi.org/10.1016/j.enpol.2014.04.032
137. Viguié, Vincent, Samuel Juhel, Tamara Ben-Ari, Morgane Colombert, James D. Ford, Louis-Gaëtan Giraudet, and Diana Reckien. 2021. "When adaptation increases energy demand: A systematic map of the literature." *Environmental Research Letters* 16(3): 033004. https://doi.org/10.1088/1748-9326/abc044

138. Fant, Charles, Brent Boehlert, Kenneth Strzepek, Peter Larsen, Alisa White, Sahil Gulati, Yue Li, and Jeremy Martinich. 2020. "Climate change impacts and costs to US electricity transmission and distribution infrastructure." *Energy* 195: 116899. https://doi.org/10.1016/j.energy.2020.116899; Voisin, Nathalie, Ana Dyreson, Tao Fu, Matt O'Connell, Sean W. D. Turner, Tian Zhou, and Jordan Macknick. 2020. "Impact of climate change on water availability and its propagation through the Western US power grid." *Applied Energy* 276: 115467. https://doi.org/10.1016/j.apenergy.2020.115467
139. Brown, Patrick R., Pieter J. Gagnon, J. Sean Corcoran, and Wesley J. Cole. 2022. "Retail rate projections for long-term electricity system models." No. NREL/TP-6A20-78224. Golden, CO: National Renewable Energy Lab.
140. Markolf, Samuel A., Christopher Hoehne, Andrew Fraser, Mikhail V. Chester, and B. Shane Underwood. 2019. "Transportation resilience to climate change and extreme weather events—beyond risk and robustness." *Transport Policy* 74: 174–186. https://doi.org/10.1016/j.tranpol.2018.11.003
141. Neumann, James E., Paul Chinowsky, Jacob Helman, Margaret Black, Charles Fant, Kenneth Strzepek, and Jeremy Martinich. 2021. "Climate effects on US infrastructure: The economics of adaptation for rail, roads, and coastal development." *Climatic Change* 167(3): 44. https://doi.org/10.1007/s10584-021-03179-w
142. Martinich, Jeremy, and Allison Crimmins. 2019. "Climate damages and adaptation potential across diverse sectors of the United States." *Nature Climate Change* 9(5): 397–404. https://doi.org/10.1038/s41558-019-0444-6
143. Underwood, B. Shane, Zack Guido, Padmini Gudipudi, and Yarden Feinberg. 2017. "Increased costs to US pavement infrastructure from future temperature rise." *Nature Climate Change* 7(10): 704–707. https://doi.org/10.1038/nclimate3390
144. Environmental Protection Agency (EPA). 2021. *Climate Change and Social Vulnerability in the United States: A Focus on Six Impacts*. EPA 430-R-21-003. Washington, D.C.: Environmental Protection Agency; Martello, Michael V., Andrew J. Whittle, Jesse M. Keenan, and Frederick P. Salvucci. 2021. "Evaluation of climate change resilience for Boston's rail rapid transit network." *Transportation Research Part D: Transport and Environment* 97: 102908. https://doi.org/10.1016/j.trd.2021.102908
145. Jacobs, Jennifer M., Lia R. Cattaneo, William Sweet, and Theodore Mansfield. 2018. "Recent and future outlooks for nuisance flooding impacts on roadways on the US East Coast." *Transportation Research Record* 2672(2): 1–10. https://doi.org/10.1177/0361198118756366
146. Jacobs et al. 2018.
147. Martinich and Crimmins 2019.
148. Brown, Thomas C., Vinod Mahat, and Jorge A. Ramirez. 2019. "Adaptation to future water shortages in the United States caused by population growth and climate change." *Earth's Future* 7(3): 219–234. https://doi.org/10.1029/2018EF001091
149. GHD. 2022. "Aquanomics: Southwestern U.S.: Home of the Megadrought." Melbourne, AU. GHD. https://aquanomics.ghd.com/en/swus.html
150. Lee County, Florida. 2023, May. "Ian progress report." https://ianprogress.leegov.com/pages/dashboards
151. Arias, Sabrina B., and Christopher W. Blair. 2023. "In the eye of the storm: Hurricanes, climate migration, and climate attitudes." *American Political Science Review* 118(4): 1593–1613. https://doi.org/10.1017/S0003055424000352
152. Kaye, Danielle, and Marisa Penaloza. 2022, November 19th. "'Your whole life is gone': Elderly retirees in Florida struggle to rebuild after Ian." National Public Radio. https://www.npr.org/2022/11/19/1137413838/your-whole-life-is-gone-elderly-retirees-in-florida-struggle-to-rebuild-after-ia
153. Guerra, John. 2023, June 1st. "In Hurricane Ian's wake, many local residents are relocating inland." *Gulfshore Business Magazine*. https://www.gulfshorebusiness.com/in-hurricane-ians-wake-many-local-residents-are-relocating-inland/
154. Reid, Robert L. 2024, July 1st. "How Florida's Babcock Ranch survived Hurricane Ian." *Civil Engineering Magazine*. https://www.asce.org/publications-and-news/civil-engineering-source/civil-engineering-magazine/issues/magazine-issue/article/2024/07/how-floridas-babcock-ranch-survived-hurricane-ian
155. U.S. Census Bureau. 2024, March 14th. Sunshine State Home to Metro Areas Among Top 10 U.S. Population Gainers From 2022 to 2023. Washington, D.C.: U.S. Census Bureau.

156. Keates, Nancy. 2023, October 20th. "Floridian are flocking to this hurricane 'safe haven.' *The Wall Street Journal*. https://www.wsj.com/story/floridians-are-flocking-to-this-hurricane-safe-haven-dc453726
157. Solomon, Adina. 2024, August 19th. "Could climate change boost 'haven' cities? *Worth Magazine*. https://worth.com/could-climate-change-boost-haven-cities/
158. Callahan, Joe. 2022, May 10th. "Ocala/Marion development: 50K residential units OK'd in past 5 years, most in southwest." *Ocala Star Banner*. https://www.ocala.com/story/news/2022/05/10/ocala-marion-county-florida-addresses-growth-related-concerns/9643547002/
159. Maxim, Alexandra, and Emily Grubert. 2021. "Effects of climate migration on town-to-city transitions in the United States: Proactive investments in civil infrastructure for resilience and sustainability." *Environmental Research: Infrastructure and Sustainability* 1(3): 031001. https://doi.org/10.1088/2634-4505/ac33ef
160. Maxim and Grubert. 2021.
161. Entergy. 2018. "Entergy Louisiana's Katrina and Rita restoration costs are paid in full." New Orleans, LA: Entergy New Orleans. https://www.entergynewsroom.com/news/entergy-louisianakatrina-rita-restoration-costs-are-paid-full/
162. Eaglesham, Jean. 2023, June 24th. "Car insurance rates are soaring with little relief in sight." *The Wall Street Journal*. https://www.wsj.com/articles/car-insurance-rates-are-soaring-with-little-relief-in-sight-66138e2a
163. Progressive Casualty Insurance Company. 2022, December 21st. "Progressive announces updated loss estimates from Hurricane Ian." https://investors.progressive.com/financials/financial-news-releases/financial-news-release-details/2022/Progressive-Announces-Updated-Loss-EstimatesFrom-Hurricane-Ian/default.aspx
164. Bankrate. 2024, July. "Average cost of car insurance in July 2024." https://www.bankrate.com/insurance/car/average-cost-of-car-insurance/
165. U.S. Census Bureau. 2024. "Louisiana: Income and poverty." Washington, D.C.: U.S. Census Bureau. https://data.census.gov/profile?q=Louisiana%20Income%20and%20Poverty&g=040XX00US22
166. Insurance Information Institute. 2023. "Domestic insurance companies by state, property/casualty and life/annuities, 2022." Malvern, PA: Insurance Information Institute. https://www.iii.org/publications/a-firm-foundation-how-insurance-supports-the-economy/a-50-state-commitment/insurance-companies-by-state
167. Insurance Information Institute. 2024. "Premium taxes by state, property/casualty, life/annuity and health insurers, 2023." Malvern, PA: Insurance Information Institute. https://www.iii.org/publications/insurance-handbook/economic-and-financial-data/state-by-state
168. Keenan, Jesse M. 2018. "Regional resilience trust funds: An exploratory analysis for leveraging insurance surcharges." *Environment Systems and Decisions* 38(1): 118–139. https://doi.org/10.1007/s10669-017-9656-3
169. Interview with Lloyds America executive, conducted in-person by the author, May 8, 2013.
170. Alexander, Sophia. 2024, December 3rd. "The quiet rise of lightly regulated home insurance." *Bloomberg*. https://www.bloomberg.com/graphics/2024-home-insurance-risky-policy/
171. Karlin, Sam. 2024, January 11th. "Louisiana welcomed small insurers looking to make money fast: Then the house of cards collapsed." *New Orleans Times-Picayune*. https://www.nola.com/news/business/11-of-12-failed-louisiana-insurers-paid-money-to-affiliates/article_a24022e2-af37-11ee-b734-ff759f54bb34.html
172. Gupta, Aparna, Abena Owusu, and Jue Wang. 2023. "Assessing US insurance firms' climate change impact and response." *The Geneva Papers on Risk and Insurance-Issues and Practice* 49: 1–34. https://doi.org/10.1057/s41288-023-00297-7
173. Insurance Information Institute. 2025. "Natural catastrophe losses in the united states by peril, 2024." Malvern, PA: Insurance Information Institute. https://www.iii.org/fact-statistic/facts-statistics-us-catastrophes
174. Zawacki, Tim, Jim Bowers, and Warren Wright. 2024, March 7th. "U.S. P/C industry underwriting loss reaches 10-year high: AM Best." *Carrier Management*. https://www.carriermanagement.com/news/2024/03/07/259633.htm
175. Zawacki et al. 2024.
176. Swiss Re. 2024, March 26th. "New record of 142 natural catastrophes accumulates to USD 108 billion insured losses in 2023, finds Swiss Re Institute." Zurich: Swiss Re. https://www.swissre.com/press-release/New-record-of-142-natural-catastrophes-accumulates-to-USD-

108-billion-insured-losses-in-2023-finds-Swiss-Re-Institute/a2512914-6d3a-492e-a190-aac37feca15b

177. Feldman, Jamie. 2023, November 13th. "More insurance companies are leaving California." Kiplinger Personal Finance. https://www.kiplinger.com/personal-finance/insurance/more-insurance-companies-are-leaving-california
178. Acosta, Deborah. 2023, October 17th. "Home insurance is so high in this Florida town, residents are leaving." *The Wall Street Journal.* https://www.wsj.com/real-estate/home-insurance-is-so-high-in-this-florida-town-residents-are-leaving-bb00c96f?
179. Department of Housing and Urban Development (HUD). 2024, April 17th. "Wind or named storm insurance coverage—maximum insurance deductibles." Mortgagee Letter: 2024-05. Washington, D.C.: Department of Housing and Urban Development.
180. Department of the Treasury. 2024, March 8th. "U.S. Department of the Treasury and state insurance regulators launch coordinated effort on homeowners insurance data collection to assess the effects of climate risk on U.S. insurance markets." Press Release. Washington, D.C.: Department of the Treasury.
181. Keys, Benjamin J. and Philip Mulder. "Property insurance and disaster risk: New evidence from mortgage escrow data." Working Paper No. 32579. Cambridge, MA: National Bureau of Economic Research. https://www.nber.org/papers/w32579
182. Jordan, Miriam. 2023, September 7th. "Desantis's immigration law may affect hurricane cleanup in Florida." *The New York Times.* https://www.nytimes.com/2023/09/07/us/florida-immigration-hurricane-cleanup.html
183. Kayyem, Juliette. 2023, August 28th. "What your insurer is trying to tell you about climate change." *The Atlantic.* https://www.theatlantic.com/ideas/archive/2023/08/home-insurance-costs-wildfires-floods-weather/675141/
184. Jergler, Don. 2018, May 8th. "California commissioner ran a climate change 'stress test' on insurers." *Insurance Journal.* https://www.insurancejournal.com/news/west/2018/05/08/488649.htm
185. Boomhower, Judson, Meredith Fowlie, Jacob Gellman, and Andrew Plantinga. 2024. "How are insurance markets adapting to climate change? Risk selection and regulation in the market for homeowners insurance." Working Paper No. 32625. Cambridge, MA: National Bureau of Economic Research. https://www.nber.org/papers/w32625
186. See note 183.
187. Beam, Adam. 2023, September 28th. "Wildfire-prone California to consider new rules for property insurance pricing." Associated Press. https://apnews.com/article/california-home-insurance-wildfire-risk-premiums-cf40911606e8e4d9c7c35ca57ca733e8
188. Beam 2023.
189. Elliott, Rebecca. 2021. "Insurance and the temporality of climate ethics: Accounting for climate change in US flood insurance." *Economy and Society* 50(2) 173–195. https://doi.org/10.1080/03085147.2020.1853356
190. De Ruig, Lars T., Toon Haer, Hans de Moel, Samuel D. Brody, W.J. Wouter Botzen, Jeffrey Czajkowski, and Jeroen C. J. H. Aerts. 2022. "How the USA can benefit from risk-based premiums combined with flood protection." *Nature Climate Change* 12(11): 995–998. https://doi.org/10.1038/s41558-022-01501-7
191. De Ruig et al. 2022.
192. Wing, Oliver E. J., Nicholas Pinter, Paul D. Bates, and Carolyn Kousky. 2020. "New insights into US flood vulnerability revealed from flood insurance big data." *Nature Communications* 11(1): 1444. https://doi.org/10.1038/s41467-020-15264-2
193. Pralle, Sarah. 2019. "Drawing lines: FEMA and the politics of mapping flood zones." *Climatic Change* 152(2): 227–237. https://doi.org/10.1007/s10584-018-2287-y
194. Ge, Shan, Ammon Lam, and Ryan Lewis. 2022. "The costs of hedging disaster risk and home prices in the face of climate change." SSRN Working Paper. https://dx.doi.org/10.2139/ssrn.4192699
195. Ratnadiwakara, Dimuthu, and Buvaneshwaran Venugopal. 2023. "Climate risk perceptions and demand for flood insurance." *Financial Management* 52(2): 297–331. https://doi.org/10.1111/fima.12414
196. Citizens Property Insurance Corporation (Citizens). 2024. "Financials: Policies in force." Tallahassee, FL: Citizens Property Insurance Corporation. https://www.citizensfla.com/policies-in-force

197. U.S. Census Bureau. 2024. "State profile: Florida." Washington, D.C.: U.S. Census Bureau. https://data.census.gov/profile/Florida?g=040XX00US12
198. Webel, Baird. 2023, November 2nd. "Testimony: Statement of Baird Webel, acting section research manager before Committee on Financial Services, Subcommittee on Housing and Insurance, U.S. Housing of Representatives hearing on 'The factors influencing the high cost of insurance for consumers.'" Washington, D.C.: Congressional Research Service. https://crsreports.congress.gov/product/pdf/TE/TE10087
199. Webel 2023.
200. Rabb, William. 2023, August 21st. "A year after uproar over Demotech rating, little has changed, but questions remain." *Insurance Journal.* https://www.insurancejournal.com/news/southeast/2023/08/21/736108.htm
201. Sastry, Parinitha, Ishita Sen, and Ana-Maria Tenekedjieva. 2023. "When insurers exit: Climate losses, fragile insurers, and mortgage markets." SSRN Working Paper. https://dx.doi.org/10.2139/ssrn.4674279
202. Florida Department of Environmental Protection. 2024. "Resilient Florida program." Tallahassee: Office of Resilience and Coastal Protection, Florida Department of Environmental Protection. https://floridadep.gov/ResilientFlorida
203. Farrington, Brendan. 2024, April 30th. "Climate change would virtually disappear in Florida—at least according to state law." Associated Press. https://apnews.com/article/florida-desantis-climate-change-environment-a3bee6775476d6f3e00b8c6cd500a3b1
204. Rohrer, Gray. 2022, December 12th. "Lawmakers plan $1B for reinsurance to stabilize property insurance industry." *Florida Politics.* https://floridapolitics.com/archives/575542-lawmakers-plan-1b-for-reinsurance-to-stabilize-property-insurance-industry/
205. U.S. Senate Committee on the Budget. 2024, March 19th. "Whitehouse presses Citizens Property Insurance for answers about company's solvency." Chairman's Newsroom. Washington, D.C.: United States Senate.
206. Sastry et al. 2023.
207. Sastry et al. 2023.
208. Sastry et al. 2023.
209. Sastry et al. 2023.
210. Sastry et al. 2023.
211. Florida Office of Insurance Regulation. 2023, January 1st. *Property Insurance Stability Report.* Tallahassee: Florida Office of Insurance Regulation.
212. Florida Office of Insurance Regulation. 2024, January 1st. *Property Insurance Stability Report.* Tallahassee: Florida Office of Insurance Regulation.
213. See note 183.
214. Karlin, Sam. 2024, May 6th. "Louisiana's insurance market may be set up for more failures." *Governing.* https://www.governing.com/management-and-administration/louisianas-insurance-market-may-be-set-up-for-more-failures
215. Karlin 2024.
216. Louisiana Citizens Property Insurance Corporation (Citizens). 2024. "Property insurers in the state of Louisiana survey." Metairie: Louisiana Citizens Property Insurance Corporation. https://www.lacitizens.com/AboutUs/property-insurers-in-the-state-of-louisiana-survey
217. Cline, Sara. 2023, August 16th. "Victor of Louisiana insurance commissioner election decided after candidate withdraws." Associated Press. https://apnews.com/article/louisiana-insurance-election-d72a255da8308509be5f6c093324b378
218. Riggs, Ben. 2024, June 20th. "Editorial: Louisiana on wrong path to address insurance crisis." *New Orleans Times-Picayune.* https://www.nola.com/opinions/louisiana-insurance-storms-reform-tim-temple-session-bills/article_4f8ed196-2d05-11ef-b17e-2fc442adb612.html
219. Adelson, Jeff. 2025, March 14th. "Population fell in two-thirds of Louisiana parishes in 2024, census estimates show." *New Orleans Times-Picayune.* https://www.nola.com/news/business/census-new-orleans-other-parishes-lost-population-in-2024/article_bf2ae7ac-003f-11f0-82df-437e777a421b.html
220. Friedman, Nicole. 2024, April 10th. "The hidden costs of homeownership are skyrocketing." *The Wall Street Journal.* https://www.wsj.com/economy/housing/housing-affordability-taxes-insurance-costs-rise-bca64df1
221. Frank, Thomas. 2024, April 25th. "Leaving Louisiana: Disasters, insurance hikes could spur mass exodus." *E&E News.* https://www.eenews.net/articles/leaving-louisiana-disasters-insurance-hikes-could-spur-mass-exodus/

222. Moody's Investors Service. 2024, April 23rd. "State and local government—Louisiana: Climate risks and insurance costs pose challenges for state and local governments." New York: Moody's Investors Service.
223. Flavelle, Christoper, and Mira Rojanasakul. 2024, May 13th. "The Home insurance crunch: See what's happening in your state." *The New York Times*. https://www.nytimes.com/interactive/2024/05/13/climate/home-insurance-profit-us-states-weather.html
224. Waller, Christopher J. 2023, May 11th. "Speech: Climate change and financial stability." Washington, D.C.: Federal Reserve Board of Governors. https://www.federalreserve.gov/newsevents/speech/waller20230511a.htm
225. Ferrara, Andreas, and Patrick A. Testa. 2023. "Churches as social insurance: Oil risk and religion in the US South." *The Journal of Economic History* 83(3): 786–832. https://doi.org/10.1017/S0022050723000268
226. Keenan, Jesse M. 2019. "A climate intelligence arms race in financial markets." *Science* 365(6459): 1240–1243. https://doi.org/10.1126/science.aay8442
227. Fiedler, Tanya, Andy J. Pitman, Kate Mackenzie, Nick Wood, Christian Jakob, and Sarah E. Perkins-Kirkpatrick. 2021. "Business risk and the emergence of climate analytics." *Nature Climate Change* 11(2): 87–94. https://doi.org/10.1038/s41558-020-00984-6
228. Mankin, Justin S. 2024, January 20th. "Editorial: The people have a right to climate data." *The New York Times*. https://www.nytimes.com/2024/01/20/opinion/climate-risk-disasters-data.html
229. Condon, Madison. 2023. "Climate services: The business of physical risk." *Arizona State Law Journal* 55: 147.

Chapter 5

1. Bittle, Jake. 2024. *The Great Displacement: Climate Change and the Next American Migration*. New York: Simon & Schuster.
2. Kamin, Debra. 2023, March 10th. "Out-of-towners head to 'climate-proof Duluth.'" *The New York Times*. https://www.nytimes.com/2023/03/10/realestate/duluth-minnesota-climate-change.html
3. Kamin 2023.
4. Allen, Samantha. 2024, January 5th. "30% of Americans cite climate change as a motivator to move in 2024." *Forbes*. https://www.forbes.com/home-improvement/features/americans-moving-climate-change/
5. Frank, Thomas. 2023, February 6th. "Disasters displaced more than 3 million Americans in 2022." *Scientific American*. https://www.scientificamerican.com/article/disasters-displaced-more-than-3-million-americans-in-2022/
6. Author's calculations; Long, Jason, and Henry Siu. 2018. "Refugees from dust and shrinking land: Tracking the dust bowl migrants." *The Journal of Economic History* 78(4): 1001–1033. https://doi.org/10.1017/S0022050718000591
7. Glanz, Theresa A. 2020. "Federal land-use policy and resettlement in the Great Plains: An experiment in community development during the New Deal years, 1933–1941." Doctoral dissertation. Lincoln, University of Nebraska.
8. Bronen, Robin. 2021. "Climigration: Creating a national governance framework for climate-forced community relocation." *NYU Review of Law & Social Change* 45: 574.
9. Pinter, Nicholas. 2021. "The lost history of managed retreat and community relocation in the United States." *Elementa: Science of the Anthropocene* 9:1. https://doi.org/10.1525/elementa.2021.00036
10. Email correspondence. Greg Peterson. June 13, 2023.
11. Skidmore, Tyler A. and Jared L. Cohon. 2023. "A multicriteria decision analysis framework for developing and evaluating coastal retreat policy." *Integrated Environmental Assessment and Management* 19(1): 83–98. https://doi.org/10.1002/ieam.4662
12. Nabong, Emily C., Lauren Hocking, Aaron Opdyke, and Jeffrey P. Walters. 2023. "Decision-making factor interactions influencing climate migration: A systems-based systematic review." *Wiley Interdisciplinary Reviews: Climate Change* 14(4): e828. https://doi.org/10.1002/wcc.828
13. Cianconi, Paolo, Sophia Betrò, and Luigi Janiri. 2020. "The impact of climate change on mental health: a systematic descriptive review." *Frontiers in Psychiatry* 11: 490206. https://doi.org/10.3389/fpsyt.2020.00074
14. Alayarian, Aida. 2019. "Trauma, resilience and healthy and unhealthy forms of dissociation." *Journal of Analytical Psychology* 64(4): 587–606. https://doi.org/10.1111/1468-5922.12522

15. Email correspondence. Zach Thorp. September 19, 2023.
16. Bukvic, Anamaria. 2017. "Towards the sustainable climate change population movement: The Relocation Suitability Index." *Climate and Development* 10(4): 307–320. https://doi.org/10.1080/17565529.2017.1291407
17. Han, Chaeyeon, Uijeong Hwang, and Subhrajit Guhathakurta. 2024, June 15th. "Can Thanksgiving destinations predict climate migration patterns?" *Findings*. https://doi.org/10.32866/001c.117432
18. Carling, Jørgen, Marta Bivand Erdal, and Cathrine Talleraas. 2021. "Living in two countries: Transnational living as an alternative to migration." *Population, Space and Place* 27(5): e2471. https://doi.org/10.1002/psp.2471
19. Interview. Valerie Nelson. March 1, 2024.
20. Van Nieuwerburgh, Stijn. 2023. "The remote work revolution: Impact on real estate values and the urban environment: 2023 AREUEA presidential address." *Real Estate Economics* 51(1): 7–48. https://doi.org/10.1111/1540-6229.12422
21. Pinter, Nicholas, and James C. Rees. 2021. "Assessing managed flood retreat and community relocation in the Midwest USA." *Natural Hazards* 107: 497–518. https://doi.org/10.1007/s11069-021-04592-1
22. Ristroph, Elizaveta Barrett. 2021. "Navigating climate change adaptation assistance for communities: A case study of Newtok Village, Alaska." *Journal of Environmental Studies and Sciences* 11(3): 329–340. https://doi.org/10.1007/s13412-021-00711-3
23. Simms, Jessica, Helen L. Waller, Chris Brunet, and Pamela Jenkins. 2021. "The long goodbye on a disappearing, ancestral island: A just retreat from Isle de Jean Charles." *Journal of Environmental Studies and Sciences* 11(3): 316–328. https://doi.org/10.1007/s13412-021-00682-5
24. Keenan, Jesse M., Thomas Hill, and Anurag Gumber. 2018. "Climate gentrification: From theory to empiricism in Miami-Dade County, Florida." *Environmental Research Letters* 13(5): 054001. https://doi.org/10.1088/1748-9326/aabb32
25. Keenan et al. 2018.
26. Tedesco, Marco, Jesse M. Keenan, and Carolynne Hultquist. 2022. "Measuring, mapping, and anticipating climate gentrification in Florida: Miami and Tampa case studies." *Cities* 131: 103991. https://doi.org/10.1016/j.cities.2022.103991
27. Shokry, Galia, Isabelle Anguelovski, James J. T. Connolly, Andrew Maroko, and Hamil Pearsall. 2022. "'They didn't see it coming': Green resilience planning and vulnerability to future climate gentrification." *Housing Policy Debate* 32(1): 211–245. https://doi.org/10.1080/10511482.2021.1944269
28. Iacuri, Greg. 2024, July 27th. "Climate change is gentrifying neighborhoods. In Miami, residents fear high prices—and a lost soul." CNBC. https://www.cnbc.com/2024/07/27/climate-gentrification-fuels-higher-prices-for-longtime-miami-residents.html; Aune, Kyle T., Dean Gesch, and Genee S. Smith. 2020. "A spatial analysis of climate gentrification in Orleans Parish, Louisiana post–Hurricane Katrina." *Environmental Research* 185: 109384. https://doi.org/10.1016/j.envres.2020.109384
29. Marandi, Anna, and Kelly Leilani Main. 2021. "Vulnerable city, recipient city, or climate destination? Towards a typology of domestic climate migration impacts in US cities." *Journal of Environmental Studies and Sciences* 11: 465–480. https://doi.org/10.1007/s13412-021-00712-2
30. Marandi and Main 2021.
31. Marandi and Main 2021.
32. Best, Kelsea B., Zeynab Jouzi, Md Sariful Islam, Timothy Kirby, Rebecca Nixon, Azmal Hossan, and Richard A. Nyiawung. 2023. "Typologies of multiple vulnerabilities and climate gentrification across the East Coast of the United States." *Urban Climate* 48: 101430. https://doi.org/10.1016/j.uclim.2023.101430
33. Seaton, Iris. 2023, June 5th. "Asheville ranks on list of cities most likely to see impact of 'climate migration." *Citizen Times* (Asheville). https://www.citizen-times.com/story/news/local/2023/06/05/will-climate-change-migration-overpopulate-asheville/70281786007/
34. Interview with Asheville Resident and NOAA Civil Servant, conducted in-person by the author, May 16, 2019.
35. Seaton, Iris. 2024, April 3rd. "Asheville ranks no. 2 in the country for highest planned move-in vs. move-out ratio." *Citizen Times* (Asheville). https://www.citizen-times.com/story/news/local/2024/04/03/asheville-is-2-in-the-country-for-highest-move-in-vs-move-out-ratio/73144952007/

36. Wilson, Scott. 2023, February 10th. "Gentrification by fire." *The Washington Post*. https://www.washingtonpost.com/nation/interactive/2023/california-fires-home-prices/
37. Wilson 2023.
38. Siders, A. R., and Jesse M. Keenan. 2020. "Variables shaping coastal adaptation decisions to armor, nourish, and retreat in North Carolina." *Ocean & Coastal Management* 183: 105023. https://doi.org/10.1016/j.ocecoaman.2019.105023
39. O'Malley, Isabella, and Jennifer McDermott. 2023, August 18th. "In Hawaii, concerns over 'climate gentrification' rise after devastating Maui fires." Associated Press. https://apnews.com/article/maui-hawaii-fire-climate-gentrification-housing-displacement-aa827eabef48d2764aa58d01f7a6969c
40. U.S. Census Bureau. 2023, September 21st. "Detailed look at Native Hawaiian and other Pacific Islander groups." Washington, D.C.: U.S. Census Bureau. https://www.census.gov/library/stories/2023/09/2020-census-dhc-a-nhpi-population.html
41. Shailer, Daniel. 2024, July 14th. "Plagued by developers and rising seas, a historic Black haven embraces conservation." *Mother Jones*. https://www.motherjones.com/politics/2024/07/10-mile-charleston-south-carolina-historic-black-community-heirs-coastal-development/
42. Bhatia, Meher. 2024, January 24th. "In Ithaca, climate change is making gentrification worse." *The Nation*. https://www.thenation.com/article/environment/ithaca-new-york-climate-change-gentrification/
43. Bhatia 2024.
44. City of Norfolk. 2024, June. "§ 3.9.19 URO: Upland resilience overlay." In *Building a Better Norfolk: A Zoning Ordinance for the 21st Century*. Norfolk, VA: City of Norfolk.
45. Roy, Manoj, David Hulme, and Ferdous Jahan. 2013. "Contrasting adaptation responses by squatters and low-income tenants in Khulna, Bangladesh." *Environment and Urbanization* 25(1): 157–176. https://doi.org/10.1177/0956247813477362
46. McNamara, Dylan E., Martin D. Smith, Zachary Williams, Sathya Gopalakrishnan, and Craig E. Landry. 2024. "Policy and market forces delay real estate price declines on the US coast." *Nature Communications* 15(1): 2209. https://doi.org/10.1038/s41467-024-46548-6
47. McNamara et al. 2024.
48. McNamara et al. 2024.
49. Dundon, Leah A., and Mark Abkowitz. 2021. "Climate-induced managed retreat in the US: A review of current research." *Climate Risk Management* 33: 100337. https://doi.org/10.1016/j.crm.2021.100337
50. Kundis Craig, Robin. 2018. "Cleaning up our toxic coasts: a precautionary and human health-based approach to coastal adaptation." *Pace Environmental Law Review* 36: 1.
51. Aktürk, Gül, and Martha Lerski. 2021. "Intangible cultural heritage: A benefit to climate-displaced and host communities." *Journal of Environmental Studies and Sciences* 11(3): 305–315. https://doi.org/10.1007/s13412-021-00697-y
52. Tansey, Eira. 2015. "Archival adaptation to climate change." *Sustainability: Science, Practice and Policy* 11(2): 45–56. https://doi.org/10.1080/15487733.2015.11908146
53. Smith, Erin, Selena Ahmed, Virgil Dupuis, MaryAnn Running Crane, Margaret Eggers, Mike Pierre, Kenneth Flagg, and Carmen Byker Shanks. 2019. "Contribution of wild foods to diet, food security, and cultural values amidst climate change." *Journal of Agriculture, Food Systems, and Community Development* 9(B): 191–214. https://doi.org/10.5304/jafscd.2019.09B.011
54. Preeshl, Artemis, and Foster Johns. 2022. "The American Elizabethan accent: Sea island residents talk." *Voice and Speech Review* 16(3): 330–340. https://doi.org/10.1080/23268263.2022.2043579
55. Elkin, Rosetta S., and Jesse M. Keenan. 2018. "Retreat or rebuild: Exploring geographic retreat in humanitarian practices in coastal communities." In *Climate Change Impacts and Adaptation Strategies for Coastal Communities*, edited by Walter Leal Filho, 149–165. Cham: Springer.
56. Sax, Dov F., Katherine F. Smith, and Andrew R. Thompson. 2009. "Managed relocation: A nuanced evaluation is needed." *Trends in Ecology and Evolution* 24(9): 472. http://dx.doi.org/10.1016/j.tree.2009.05.004
57. Sax et al. 2009.
58. Bower, Erica R., Anvesh Badamikar, Gabrielle Wong-Parodi, and Christopher B. Field. 2023. "Enabling pathways for sustainable livelihoods in planned relocation." *Nature Climate Change* 13(9): 919–926. https://doi.org/10.1038/s41558-023-01753-x
59. Mueller, Noah J., and Christopher F. Meindl. 2017. "Vulnerability of Caribbean island cemeteries to sea level rise and storm surge." *Coastal Management* 45(4): 277–292. https://doi.org/10.1080/08920753.2017.1327343

60. See generally, Karla Rothstein. 2024. "Death Lab." New York: Graduate School of Architecture, Planning, and Preservation, Columbia University. https://www.arch.columbia.edu/research/labs/3-death-lab
61. Pulver, Dinah Voyles. 2023, September 19th. "When the dead don't stay buried: The grave situation at cemeteries amid climate change." *USA Today*. https://www.usatoday.com/story/news/nation/2023/09/14/climate-change-cemetery-effects/7882519001/

Chapter 6

1. Schneekloth, Lynda H. 1998. "Uredeemably utopian: Architecture and making/unmaking the world." *Utopian Studies* 9(1): 1–25.
2. Milman, Oliver. 2018, September 24th. "Where should you move to save yourself from climate change?" *The Guardian*. https://www.theguardian.com/environment/2018/sep/24/climate-change-where-to-move-us-avoid-floods-hurricanes
3. Milman 2018.
4. Teicher, Hannah M., and Patrick Marchman. 2024. "Integration as adaptation: Advancing research and practice for inclusive climate receiving communities." *Journal of the American Planning Association* 90(1): 30–49. https://doi.org/10.1080/01944363.2023.2188242
5. Winkler, Richelle L., and Mark D. Rouleau. 2021 "Amenities or disamenities? Estimating the impacts of extreme heat and wildfire on domestic US migration." *Population and Environment* 42: 622–648. https://doi.org/10.1007/s11111-020-00364-4
6. Berlin Rubin, Nina, and Gabrielle Wong-Parodi. 2022. "As California burns: The psychology of wildfire-and wildfire smoke-related migration intentions." *Population and Environment* 44(1): 15–45. https://doi.org/10.1007/s11111-022-00409-w
7. Gosnell, Hannah, and Jesse Abrams. 2011. "Amenity migration: Diverse conceptualizations of drivers, socioeconomic dimensions, and emerging challenges." *GeoJournal* 76: 303–322. https://doi.org/10.1007/s10708-009-9295-4
8. De Sherbinin, Alex, Kathryn Grace, Sonali McDermid, Kees van der Geest, Michael J. Puma, and Andrew Bell. 2022. "Migration theory in climate mobility research." *Frontiers in Climate* 4: 882343. https://doi.org/10.3389/fclim.2022.882343
9. Ton, Marijn J., Hans de Moel, Jens A. de Bruijn, Wouter J. W. Botzen, Hande Karabiyik, Marina Friedrich, and Jeroen Aerts. 2024. "The impact of natural hazards on migration in the United States and the effect of spatial dependence." *Journal of Environmental Planning and Management* 1–19. https://doi.org/10.1080/09640568.2024.2359447
10. Donnelly, Sean. 2023, October 25th. "Most climate-resilient cities (2024)." *Architectural Digest*. Click here to enter text.
11. Sutter, John. 2021, April 12th. "As people flee climate change on the coasts, this Midwest city is trying to become a safe haven." CNN. https://www.cnn.com/2021/04/12/opinions/climate-migration-in-america-california-duluth-sutter/index.html
12. Milwaukee Partnership for Economic Development. 2024. "Location benefits." https://mkeregion.com/location/
13. Be in Buffalo. 2024. "Buffalo Niagara as a climate change refuge." https://beinbuffalo.com/community/climate-refuge/
14. Be in Buffalo 2024.
15. McDaniel, Paul N., Darlene Xiomara Rodriguez, and Qingfang Wang. 2019. "Immigrant integration and receptivity policy formation in welcoming cities." *Journal of Urban Affairs* 41(8): 1142–1166. https://doi.org/10.1080/07352166.2019.1572456
16. De Socio, Mike. 2021, April 19th. "The problem with 'climate havens.'" *Bloomberg CityLab*. https://www.bloomberg.com/news/articles/2021-04-19/there-s-no-such-thing-as-a-climate-haven
17. De Socio 2021.
18. De Socio, Mike. 2024, July 1st. "US cities are advertising themselves as 'climate havens': But can they actually protect residents from extreme weather?" BBC. https://www.bbc.com/future/article/20240628-us-climate-havens-cities-claim-extreme-weather-protection
19. Ashik, Fajle Rabbi, M. H. Rahman, and M. Kamruzzaman. 2022. "Investigating the impacts of transit-oriented development on transport-related CO2 emissions." *Transportation Research Part D: Transport and Environment* 105: 103227. https://doi.org/10.1016/j.trd.2022.103227

20. Cervero, Robert, and Cathleen Sullivan. 2011. "Green TODs: marrying transit-oriented development and green urbanism." *International Journal of Sustainable Development & World Ecology* 18(3): 210–218. https://doi.org/10.1080/13504509.2011.570801
21. Marino, E. K., K. Maxwell, E. Eisenhauer, A. Zycherman, C. Callison, E. Fussell, M. D. Hendricks, F. H. Jacobs, A. Jerolleman, A. K. Jorgenson, E. M. Markowitz, S. T. Marquart-Pyatt, M. Schutten, R. L. Shwom, and K. Whyte. 2023. "Ch. 20: Social Systems and Justice." In Crimmins et al., *Fifth National Climate Assessment.* https://doi.org/10.7930/NCA5.2023.CH20
22. De Socio 2024.
23. Dewey, Caitlin. 2018, November 24th. "Hurricane refugees liked what they found in Buffalo—so they're staying." *The Buffalo News.* https://buffalonews.com/news/local/hurricane-refugees-liked-what-they-found-in-buffalo-so-theyre-staying/article_c69708c3-fc0a-583f-a8ce-c8f2060c514d.html
24. Yoder, Kate. 2021, December 7th. "Fleeing global warming? 'Climate havens' aren't ready for you yet." *Grist.* https://grist.org/migration/fleeing-global-warming-climate-havens-arent-ready-for-you-yet/
25. Maxim, Alexandra, and Emily Grubert. 2022. "Anticipating climate-related changes to residential energy burden in the United States: Advance planning for equity and resilience." *Environmental Justice* 15(3): 139–148. https://doi.org/10.1089/env.2021.0056
26. Davidson, Kyle. 2023, August 4th. "Health and climate justice experts caution against labeling Michigan a 'climate haven.'" *Michigan Advance.* https://michiganadvance.com/2023/08/05/health-and-climate-justice-experts-caution-against-labeling-michigan-a-climate-haven/
27. Fussell, Elizabeth, Jack DeWaard, and Katherine J. Curtis. 2023. "Environmental migration as short- or long-term differences from a trend: A case study of Hurricanes Katrina and Rita effects on out-migration in the Gulf of Mexico." *International Migration* 61(5): 60–74. https://doi.org/10.1111/imig.13101
28. Fussell, Elizabeth, Katherine J. Curtis, and Jack DeWaard. 2014. "Recovery migration to the City of New Orleans after Hurricane Katrina: A migration systems approach." *Population and Environment* 35: 305–322. https://doi.org/10.1007/s11111-014-0204-5
29. Marandi and Main 2021.
30. Arbit, Julie, Brad Bottoms, and Earl Lewis. 2023, August 26th. "Why these 'climate haven' cities aren't yet ready for more extreme weather events." PBS. https://www.pbs.org/newshour/science/why-these-climate-haven-cities-arent-yet-ready-for-more-extreme-weather-events
31. Email correspondence. Keegan Pyle. June 25, 2023.
32. Abrams, Samuel J., and Morris P. Fiorina. 2012. "'The big sort' that wasn't: A skeptical reexamination." *PS: Political Science & Politics* 45(2): 203–210. https://doi.org/10.1017/S1049096512000017
33. Stossel, Scott. 2008, May 18th. "Subdivided we fall." *The New York Times.* https://www.nytimes.com/2008/05/18/books/review/Stossel-t.html
34. Morris, Erin, Joshua J. Cousins, and Andrea Feldpausch-Parker. 2023. "Transformation and recognition: Planning just climate havens in New York State." *Environmental Science & Policy* 146: 57–65. https://doi.org/10.1016/j.envsci.2023.05.004
35. Morris et al. 2023.
36. Morris et al. 2023.
37. Pierre-Louis, Kendra. 2019, April 15th. "Want to escape global warming? These cities promise cool relief." *The New York Times.* https://www.nytimes.com/2019/04/15/climate/climate-migration-duluth.html
38. Keenan, Jesse M. 2019. "Destination Duluth: Competitive economic development in the age of climigration." In *Proceedings of Global Shifts: A Changing Climate, A Changing World*: 1–3. Philadelphia: University of Pennsylvania, Perry World House.
39. Sutter, John. 2021, April 21st. "People are moving to this Midwest city for one particular reason." CNN. https://youtu.be/lbQQtOJvSkA?si=GeBtaYBStbJN8Duc
40. Kosta, Michael. 2024, February 28th. "Is Duluth, MN a climate safe haven?" *The Daily Show* (Comedy Central). https://www.youtube.com/watch?v=83DX5XDwz40
41. *Extrapolations.* 2023. "The fifth question." Apple TV+. https://www.apple.com/tv-pr/originals/extrapolations/
42. Kamin, Debra. 2023, March 13th. "Out-of-towners head to 'climate-proof Duluth.'" *The New York Times.* https://www.nytimes.com/2023/03/10/realestate/duluth-minnesota-climate-change.html

43. Kamin 2023.
44. Hemphill, Stephanie. 2009, February 27th. "Duluth businesses test sustainable approach." Minnesota Public Radio. https://www.mprnews.org/story/2009/02/27/duluth-businesses-test-sustainable-approach
45. Sterner, Robert W., Bonnie Keeler, Stephen Polasky, Rajendra Poudel, Kirsten Rhude, and Maggie Rogers. 2020. "Ecosystem services of earth's largest freshwater lakes." *Ecosystem Services* 41: 101046. https://doi.org/10.1016/j.ecoser.2019.101046
46. Keenan, Jesse M. 2019, March 20th. "Destination Duluth: The fact and fiction of a shared climate future." Lecture delivered at the University of Minnesota, Duluth. Retrieved from https://youtu.be/vUhSckQsrwY?si=5T7s1FpFg3jnb1AP
47. The phrase "Climate-Proof Duluth" was offered as a moment of humorous reflection, as no place can be climate-proof. Unfortunately, this became a distracting point of misinformation.
48. See note 42; Riley, Hunter. 2024, April 23rd. "'Climigrants' seek refuge in the 'air-conditioned city.'" *Lake Voice.* https://lakevoicenews.org/climagrants-seek-refuge-in-the-air-conditioned-city-733580f16729
49. Kaplan, Michael. 2024, January 23rd. "Locals 'worried' as Minnesota billionaire massively overspends on 'piece of crap' houses." *New York Post.* https://nypost.com/2024/01/23/news/locals-nervous-as-minnesota-billionaire-buys-10-crap-houses/
50. See note 42.
51. Riley 2024.
52. Barkemeyer, Ralf, Diane Holt, Lutz Preuss, and Stephen Tsang. 2014. "What happened to the 'development' in sustainable development? Business guidelines two decades after Brundtland." *Sustainable Development* 22(1): 15–32. https://doi.org/10.1002/sd.521
53. Norman, Wayne, and Chris MacDonald. 2004. "Getting to the bottom of 'triple bottom line.'" *Business Ethics Quarterly* 14(2): 243–262. https://doi.org/10.5840/beq200414211
54. Vos, Robert O. 2007. "Defining sustainability: A conceptual orientation." *Journal of Chemical Technology and Biotechnology* 82(4): 334–339. https://doi.org/10.1002/jctb.1675
55. Pearce, Annie R., and Yong Han Ahn. 2013. *Sustainable Buildings and Infrastructure: Paths to the Future.* New York: Routledge.
56. Belmonte-Ureña, Luis Jesús, José Antonio Plaza-Úbeda, Diego Vazquez-Brust, and Natalia Yakovleva. 2021. "Circular economy, degrowth and green growth as pathways for research on sustainable development goals: A global analysis and future agenda." *Ecological Economics* 185: 107050. https://doi.org/10.1016/j.ecolecon.2021.107050
57. Belmonte-Ureña et al. 2021.
58. National Institute of Standards and Technology (NIST). 2024. "Building for environmental and economic sustainability (BEES)." Online Version 2.1. Gaithersburg, MD: Department of Commerce. https://ws680.nist.gov/Bees2
59. Franconi, Ellen, Jeremy Lerond, Chitra Nambiar, Dongsu Kim, David Winiarski, Michael Rosenberg, and Yunyang Ye. 2020. *Filling the Efficiency Gap to Achieve Zero-Energy Buildings with Energy Codes.* Oakridge, TN: Pacific Northwest National Laboratory, Department of Energy.
60. Janjua, Shahana Y., Prabir K. Sarker, and Wahidul K. Biswas. 2020. "Development of triple bottom line indicators for life cycle sustainability assessment of residential buildings." *Journal of Environmental Management* 264: 110476. https://doi.org/10.1016/j.jenvman.2020.110476
61. Brown, Marilyn A., Anmol Soni, Melissa V. Lapsa, Katie Southworth, and Matt Cox. 2020. "High energy burden and low-income energy affordability: Conclusions from a literature review." *Progress in Energy* 2(4): 042003. https://doi.org10.1088/2516-1083/abb954
62. Bell, Michael, and Eunjeong Seong. 2025. *8 Minutes, 20 Seconds.* New York: Actar.
63. Jin, Hongyu, Melissa Chan, Romana Morda, Catherine Xiaocui Lou, and Zora Vrcelj. 2023. "A scientometric review of sustainable infrastructure research: Visualization and analysis." *International Journal of Construction Management* 23(11): 1847–1855. https://doi.org/10.1080/15623599.2021.2017114
64. Hoover, Fushcia-Ann, Sara Meerow, Emma Coleman, Zbigniew Grabowski, and Timon McPhearson. 2023. "Why go green? Comparing rationales and planning criteria for green infrastructure in US city plans." *Landscape and Urban Planning* 237: 104781. https://doi.org/10.1016/j.landurbplan.2023.104781
65. Maxim, Alexandra, and Emily Grubert. 2021. "Effects of climate migration on town-to-city transitions in the United States: Proactive investments in civil infrastructure for resilience

and sustainability." *Environmental Research: Infrastructure and Sustainability* 1(3): 031001. https://doi.org/10.1088/2634-4505/ac33ef

66. Forsyth, Ann, and Richard Peiser. 2021. "Lessons from planned resettlement and new town experiences for avoiding climate sprawl." *Landscape and Urban Planning* 205: 103957. https://doi.org/10.1016/j.landurbplan.2020.103957
67. Forsyth and Peiser 2021.
68. Forsyth and Peiser 2021.
69. Scott, Mark, Menelaos Gkartzios, and Keith Halfacree. 2024. "Introducing climate-related counterurbanisation: Individual adaptation or societal maladaptation?" *Habitat International* 143: 102970. https://doi.org/10.1016/j.habitatint.2023.102970
70. Brigaud, Millie. 2023, September 15th. "They moved to Vermont for climate safety: Then came floods." *Christian Science Monitor.* https://www.csmonitor.com/Environment/2023/0915/They-moved-to-Vermont-for-climate-safety.-Then-came-floods
71. Daniels, Thomas L., Kyle McCarthy, and Mark B. Lapping. 2023. "The fragmenting countryside and the challenge of retaining agricultural land: The Vermont case." *Society & Natural Resources* 36(1): 40–57. https://doi.org/10.1080/08941920.2022.2132438
72. See note 62.
73. Hurlbut, David J., Jianyu Gu, Srihari Sundar, An Pham, Barbara O'Neill, Heather Buchanan, Donna Heimiller, Mark Weimar, and Kyle Wilson. 2024. *Interregional Renewable Energy Zones.* NREL/TP-6A20-88228. Golden, CO: National Renewable Energy Laboratory. https://www.nrel.gov/docs/fy24osti/88228.pdf

Index

For the benefit of digital users, indexed terms that span two pages (e.g., 52–53) may, on occasion, appear on only one of those pages.

Note: Page numbers followed by *f* and *t* indicate figures (and maps), and tables, respectively. Numbers followed by n indicate endnotes.

G

X

Y

Z